HANDBOOK OF CARBON, GRAPHITE, DIAMOND AND FULLERENES

HANDBOOK OF CARBON, GRAPHITE, DIAMOND AND FULLERENES

Properties, Processing and Applications

by

Hugh O. Pierson

Consultant and Sandia National Laboratories (retired)
Albuquerque, New Mexico

np **NOYES PUBLICATIONS**
Park Ridge, New Jersey, U.S.A.

Library of Congress Catalog Card Number: 93-29744
ISBN: 0-8155-1339-9
Printed in the United States

Published in the United States of America by
Noyes Publications
Mill Road, Park Ridge, New Jersey 07656

Library of Congress Cataloging-in-Publication Data

Pierson, Hugh O.
    Handbook of carbon, graphite, diamond, and fullerenes : properties,
 processing, and applications / by Hugh O. Pierson.
          p.   cm.
    Includes bibliographical references and index.
    ISBN 0-8155-1339-9
    1. Carbon.   I. Title
 TP245.C4P54   1993
 661' .0681--dc20   93-29744
                                           CIP

Transferred to Digital Printing in 2009.

MATERIALS SCIENCE AND PROCESS TECHNOLOGY SERIES

Editors

Rointan F. Bunshah, University of California, Los Angeles *(Series Editor)*
Gary E. McGuire, Microelectronics Center of North Carolina *(Series Editor)*
Stephen M. Rossnagel, IBM Thomas J. Watson Research Center
(Consulting Editor)

Electronic Materials and Process Technology

HANDBOOK OF DEPOSITION TECHNOLOGIES FOR FILMS AND COATINGS, Second Edition: edited by Rointan F. Bunshah

CHEMICAL VAPOR DEPOSITION FOR MICROELECTRONICS: by Arthur Sherman

SEMICONDUCTOR MATERIALS AND PROCESS TECHNOLOGY HANDBOOK: edited by Gary E. McGuire

HYBRID MICROCIRCUIT TECHNOLOGY HANDBOOK: by James J. Licari and Leonard R. Enlow

HANDBOOK OF THIN FILM DEPOSITION PROCESSES AND TECHNIQUES: edited by Klaus K. Schuegraf

IONIZED-CLUSTER BEAM DEPOSITION AND EPITAXY: by Toshinori Takagi

DIFFUSION PHENOMENA IN THIN FILMS AND MICROELECTRONIC MATERIALS: edited by Devendra Gupta and Paul S. Ho

HANDBOOK OF CONTAMINATION CONTROL IN MICROELECTRONICS: edited by Donald L. Tolliver

HANDBOOK OF ION BEAM PROCESSING TECHNOLOGY: edited by Jerome J. Cuomo, Stephen M. Rossnagel, and Harold R. Kaufman

CHARACTERIZATION OF SEMICONDUCTOR MATERIALS, Volume 1: edited by Gary E. McGuire

HANDBOOK OF PLASMA PROCESSING TECHNOLOGY: edited by Stephen M. Rossnagel, Jerome J. Cuomo, and William D. Westwood

HANDBOOK OF SEMICONDUCTOR SILICON TECHNOLOGY: edited by William C. O'Mara, Robert B. Herring, and Lee P. Hunt

HANDBOOK OF POLYMER COATINGS FOR ELECTRONICS, 2nd Edition: by James Licari and Laura A. Hughes

HANDBOOK OF SPUTTER DEPOSITION TECHNOLOGY: by Kiyotaka Wasa and Shigeru Hayakawa

HANDBOOK OF VLSI MICROLITHOGRAPHY: edited by William B. Glendinning and John N. Helbert

CHEMISTRY OF SUPERCONDUCTOR MATERIALS: edited by Terrell A. Vanderah

CHEMICAL VAPOR DEPOSITION OF TUNGSTEN AND TUNGSTEN SILICIDES: by John E. J. Schmitz

ELECTROCHEMISTRY OF SEMICONDUCTORS AND ELECTRONICS: edited by John McHardy and Frank Ludwig

v

HANDBOOK OF CHEMICAL VAPOR DEPOSITION: by Hugh O. Pierson

DIAMOND FILMS AND COATINGS: edited by Robert F. Davis

ELECTRODEPOSITION: by Jack W. Dini

HANDBOOK OF SEMICONDUCTOR WAFER CLEANING TECHNOLOGY: edited by Werner Kern

CONTACTS TO SEMICONDUCTORS: edited by Leonard J. Brillson

HANDBOOK OF MULTILEVEL METALLIZATION FOR INTEGRATED CIRCUITS: edited by Syd R. Wilson, Clarence J. Tracy, and John L. Freeman, Jr.

HANDBOOK OF CARBON, GRAPHITE, DIAMONDS AND FULLERENES: by Hugh O. Pierson

Ceramic and Other Materials—Processing and Technology

SOL-GEL TECHNOLOGY FOR THIN FILMS, FIBERS, PREFORMS, ELECTRONICS AND SPECIALTY SHAPES: edited by Lisa C. Klein

FIBER REINFORCED CERAMIC COMPOSITES: edited by K. S. Mazdiyasni

ADVANCED CERAMIC PROCESSING AND TECHNOLOGY, Volume 1: edited by Jon G. P. Binner

FRICTION AND WEAR TRANSITIONS OF MATERIALS: by Peter J. Blau

SHOCK WAVES FOR INDUSTRIAL APPLICATIONS: edited by Lawrence E. Murr

SPECIAL MELTING AND PROCESSING TECHNOLOGIES: edited by G. K. Bhat

CORROSION OF GLASS, CERAMICS AND CERAMIC SUPERCONDUCTORS: edited by David E. Clark and Bruce K. Zoitos

HANDBOOK OF INDUSTRIAL REFRACTORIES TECHNOLOGY: by Stephen C. Carniglia and Gordon L. Barna

CERAMIC FILMS AND COATINGS: edited by John B. Wachtman and Richard A. Haber

Related Titles

ADHESIVES TECHNOLOGY HANDBOOK: by Arthur H. Landrock

HANDBOOK OF THERMOSET PLASTICS: edited by Sidney H. Goodman

SURFACE PREPARATION TECHNIQUES FOR ADHESIVE BONDING: by Raymond F. Wegman

FORMULATING PLASTICS AND ELASTOMERS BY COMPUTER: by Ralph D. Hermansen

HANDBOOK OF ADHESIVE BONDED STRUCTURAL REPAIR: by Raymond F. Wegman and Thomas R. Tullos

CARBON–CARBON MATERIALS AND COMPOSITES: edited by John D. Buckley and Dan D. Edie

CODE COMPLIANCE FOR ADVANCED TECHNOLOGY FACILITIES: by William R. Acorn

Foreword

To say that carbon is a unique element is perhaps self-evident. All elements are unique, but carbon especially so. Its polymorphs range from the hard, transparent diamond to the soft, black graphite, with a host of semi-crystalline and amorphous forms also available. It is the only element which gives its name to two scientific journals, *Carbon* (English) and *Tanso* (Japanese). Indeed, I do not know of another element which can claim to name *one* journal.

While there have been recent books on specific forms of carbon notably carbon fibers, it is a long time since somebody had the courage to write a book which encompassed all carbon materials. High Pierson perhaps did not know what he was getting into when he started this work. The recent and ongoing research activity on diamond-like films and the fullerenes, both buckyballs and buckytubes, has provided, almost daily, new results which, any author knows, makes an attempt to cover them almost futile.

In this book, the author provides a valuable, up-to-date account of both the newer and traditional forms of carbon, both naturally occurring and man-made.

An initial reading of chapters dealing with some very familiar and some not-so-familiar topics, shows that the author has make an excellent attempt to cover the field. This volume will be a valuable resource for both specialists in, and occasional users of, carbon materials for the foreseeable future. I am delighted to have had the opportunity to see the initial manuscript and to write this foreword.

Peter A. Thrower
Editor-in-Chief, *CARBON*

Preface

This book is a review of the science and technology of the element carbon and its allotropes: graphite, diamond and the fullerenes. This field has expanded greatly in the last three decades stimulated by many major discoveries such as carbon fibers, low-pressure diamond and the fullerenes. The need for such a book has been felt for some time.

These carbon materials are very different in structure and properties. Some are very old (charcoal), others brand new (the fullerenes). They have different applications and markets and are produced by different segments of the industry. Yet they have a common building block: the element carbon which bonds the various sections of the book together.

The carbon and graphite industry is in a state of considerable flux as new designs, new products and new materials, such as high-strength fibers, glassy carbon and pyrolytic graphite, are continuously being introduced.

Likewise, a revolution in the diamond business is in progress as the low-pressure process becomes an industrial reality. It will soon be possible to take advantage of the outstanding properties of diamond to develop a myriad of new applications. The production of large diamond crystal at low cost is a distinct possibility in the not-too-distant future and may lead to a drastic change of the existing business structure.

The fullerenes may also create their own revolution in the development of an entirely new branch of organic chemistry.

For many years as head of the Chemical Vapor Deposition laboratory and a contributor to the carbon-carbon program at Sandia National Laboratories and now as a consultant, I have had the opportunity to review and study the many aspects of carbon and diamond, their chemistry, technology, processes, equipment and applications, that provide the necessary background for this book.

I am indebted to an old friend, Arthur Mullendore, retired from Sandia National Laboratories, for his many ideas, comments and thorough review of the manuscript. I also wish to thank the many people who helped in the preparation and review of the manuscript and especially Peter Thrower, Professor at Pennsylvania State University and editor of *Carbon;* William Nystrom, Carbone-Lorraine America; Walter Yarborough, Professor at Pennsylvania State University; Thomas Anthony, GE Corporate Research and Development; Gus Mullen and Charles Logan, BP Chemicals: Rithia Williams, Rocketdyne. Thanks also to Bonnie Skinendore for preparing the illustrations, and to George Narita, executive editor of Noyes Publications, for his help and patience.

September 1993   Hugh O. Pierson
Albuquerque, New Mexico

Contents

9   Applications of Carbon Fibers 198

1

Introduction and General Considerations

1.0 BOOK OBJECTIVES

Many books and reviews have been published on the subject of carbon, each dealing with a specific aspect of the technology, such as carbon chemistry, graphite fibers, carbon activation, carbon and graphite properties, and the many aspects of diamond.

However few studies are available that attempt to review the entire field of carbon as a whole discipline. Moreover these studies were written several decades ago and are generally outdated since the development of the technology is moving very rapidly and the scope of applications is constantly expanding and reaching into new fields such as aerospace, automotive, semiconductors, optics and electronics.

The author and some of his colleagues felt the need for an updated and systematic review of carbon and its allotropes which would summarize the scientific and engineering aspects, coordinate the divergent trends found today in industry and the academic community, and sharpen the focus of research and development by promoting interaction. These are the objectives of this book

2.0 THE CARBON ELEMENT AND ITS VARIOUS FORMS

2.1 The Element Carbon

The word *carbon* is derived from the Latin "carbo", which to the Romans meant charcoal (or ember). In the modern world, carbon is, of course, much more than charcoal. From carbon come the highest strength fibers, one of the best lubricants (graphite), the strongest crystal and hardest material (diamond), an essentially non-crystalline product (vitreous carbon), one of the best gas adsorbers (activated charcoal), and one of the best helium gas barriers (vitreous carbon). A great deal is yet to be learned and new forms of carbon are still being discovered such as the fullerene molecules and the hexagonal polytypes of diamond.

These very diverse materials, with such large differences in properties, all have the same building block—the element carbon—which is the thread that ties the various constituents of this book and gives it unity.

2.2 Carbon Terminology

The carbon terminology can be confusing because carbon is different from other elements in one important respect, that is its diversity. Unlike most elements, carbon has several material forms which are known as polymorphs (or allotropes). They are composed entirely of carbon but have different physical structures and, uniquely to carbon, have different names: graphite, diamond, lonsdalite, fullerene, and others.

In order to clarify the terminology, it is necessary to define what is meant by carbon and its polymorphs. When used by itself, the term "carbon" should only mean the element. To describe a "carbon" material, the term is used with a qualifier such as carbon fiber, pyrolytic carbon, vitreous carbon, and others. These carbon materials have an sp^2 atomic structure, and are essentially graphitic in nature.

Other materials with an sp^3 atomic structure are, by common practice, called by the name of their allotropic form, i.e., diamond, lonsdalite, etc., and not commonly referred to as "carbon" materials, although, strictly speaking, they are.

The presently accepted definition of these words, carbon, graphite, diamond, and related terms, is given in the relevant chapters. These definitions are in accordance with the guidelines established by the *International Committee for Characterization and Terminology of Carbon* and regularly published in the journal *Carbon*.

2.3 Carbon and Organic Chemistry

The carbon element is the basic constituent of all organic matter and the key element of the compounds that form the huge and very complex discipline of organic chemistry. However the focus of this book is the polymorphs of carbon and not its compounds, and only those organic compounds that are used as precursors will be reviewed.

3.0 THE CARBON ELEMENT IN NATURE

3.1 The Element Carbon on Earth

The element carbon is widely distributed in nature.[1] It is found in the earth's crust in the ratio of 180 ppm, most of it in the form of compounds.[2] Many of these natural compounds are essential to the production of synthetic carbon materials and include various coals (bituminous and anthracite), hydrocarbons complexes (petroleum, tar, and asphalt) and the gaseous hydrocarbons (methane and others).

Only two polymorphs of carbon are found on earth as minerals: natural graphite (reviewed in Ch. 10) and diamond (reviewed in Chs. 11 and 12).

3.2 The Element Carbon in the Universe

The element carbon is detected in abundance in the universe, in the sun, stars, comets, and in the atmosphere of the planets. It is the fourth most abundant element in the solar system, after hydrogen, helium, and oxygen, and is found mostly in the form of hydrocarbons and other compounds. The spontaneous generation of fullerene molecules may also play an important role in the process of stellar dust formation.[3] Carbon polymorphs, such as microscopic diamond and *lonsdaleite,* a form similar to diamond, have been discovered in some meteorites (see Ch. 11).[4]

4.0 HISTORICAL PERSPECTIVE

Carbon, in the form of charcoal, is an element of prehistoric discovery and was familiar to many ancient civilizations. As diamond, it has been known since the early history of mankind. A historical perspective of carbon and its allotropes and the important dates in the development of carbon technology are given in Table 1.1. Additional notes of historical interest will be presented in the relevant chapters.

Table 1.1. Chronology of Carbon

First "lead" pencils	1600's
Discovery of the carbon composition of diamond	1797
First carbon electrode for electric arc	1800
Graphite recognized as a carbon polymorph	1855
First carbon filament	1879
Chemical vapor deposition (CVD) of carbon patented	1880
Production of first molded graphite (Acheson process)	1896
Carbon dating with ^{14}C isotope	1946
Industrial production of pyrolytic graphite	1950's
Industrial production of carbon fibers from rayon	1950's
Development and production of vitreous carbon	1960's
Development of PAN-based carbon fibers	1960's
Development of pitch-based carbon fibers	late 1960's
Discovery of low-pressure diamond synthesis	1970's
Production of synthetic diamond suitable for gem trade	1985
Development of diamond-like carbon (DLC)	1980's
Discovery of the fullerene molecules	late 1980's
Industrial production of CVD diamond	1992

5.0 PRODUCTS DERIVED FROM THE CARBON ELEMENT

5.1 Typical Examples

Products derived from the carbon element are found in most facets of everyday life, from the grimy soot in the chimney to the diamonds in the jewelry box. They have an extraordinary broad range of applications, illustrated by the following examples current in 1993.

- Natural graphite for lubricants and shoe polish
- Carbon black reinforcement essential to every automobile tire
- Carbon black and lamp black found in all printing inks
- Acetylene black in conductive rubber
- Vegetable and bone chars to decolorize and purify sugar and other food
- Activated charcoal for gas purification and catalytic support
- Carbon-carbon composites for aircraft brakes and space shuttle components
- High-strength carbon fibers for composite materials
- Very large graphite electrodes for metal processing
- Carbon black for copying machines
- Graphite brushes and contacts for electrical machinery
- Diamond optical window for spacecrafts
- Polycrystalline diamond coatings for cutting tools
- Low-pressure processed diamond heat-sinks for ultrafast semiconductors

5.2 Process and Product Classification

As mentioned above, only the minerals diamond and natural graphite are found in nature. All other carbon products are man-made and derive from carbonaceous precursors. These synthetic products are manufactured by a number of processes summarized in Table 1.2. Each process will be reviewed in the relevant chapters.

In this book, the applications of carbon materials are classified by product functions such as chemical, structural, electrical, and optical. This classification corresponds roughly to the various segments of industry including aerospace and automotive, metals and chemicals, electronics and semiconductor, optics, and photonics.

Table 1.2. Major Processes for the Production of Carbon Materials

Process	Carbon Product
Molding/carbonization	Molded graphite Vitreous carbon
Pyrolysis/combustion	Lampblack Carbon black
Extrusion/carbonization	Carbon fiber
High-pressure/shock	Diamond
Chemical Vapor Deposition	Polycrystalline diamond Pyrolytic graphite
Sputtering/plasma	Diamond-like carbon (DLC)

6.0 PROFILE OF THE INDUSTRY

6.1 Overview of the Industry

The wide variety of carbon-derived materials is reflected in the diversity of the industry, from small research laboratories developing diamond coatings to very large plants producing graphite electrodes. Together, these organizations form one of the world's major industries.

However, black art and secrecy still prevail in many sectors and progress often seems to occur independently with little interaction and coordination when actually the various technologies share the same scientific basis, the same principles, the same chemistry, and in many cases the same equipment. A purpose and focus of this book is to bring these divergent areas together in one unified whole and to accomplish, in a book form, what has been the goal for many years of several academic groups such as the Pennsylvania State University.

Yet progress is undeniable. The technology is versatile and dynamic and the scope of its applications is constantly expanding. It is significant that three of the most important discoveries in the field of materials in the last thirty years are related to carbon: carbon fibers, low-pressure diamond synthesis, and, very recently, the fullerene molecules.

6.2 Market

The market for carbon-derived products is divided into two major categories: carbon/graphite products and diamond with global markets of $5.5 billion and $7.5 billion respectively. These and the following figures are based on U.S. Government statistics and other sources and are to be regarded as broad estimates.[5] Additional details on the market will be given in the relevant chapters.

Market for Carbon and Graphite Products. Table 1.3 lists the estimated markets for the various forms of carbon and graphite reviewed in Chs. 5 to 10. The old and well-established industry of molded carbon and graphite still has a major share of the market but the market for others such as carbon fibers is expanding rapidly.

Table 1.3. Estimated World Market for Carbon and Graphite Products in 1991

	$ million
Molded carbon and graphite	3740
Polymeric carbon, vitreous carbon and foam	30
Pyrolytic graphite	30
Carbon fibers	200
Carbon fiber composites	700
Carbon and graphite particles and powders	800
Total	5500

Market for Diamond Products. Table 1.4 gives an estimate of the market for the various categories of diamond.

Gemstones, with over 90% of the market, still remain the major use of diamond from a monetary standpoint, in a business tightly controlled by a worldwide cartel dominated by the de Beers Organization of South Africa. The industrial diamond market is divided between natural and high-pressure synthetic diamond, the latter having the larger share of the market. This market includes coatings of CVD diamond and diamond-like carbon (DLC) which have a small but rapidly-growing share.

Table 1.4. Estimated World Market for Diamond Products in 1991

		$ million
Gemstones		7000
Industrial diamonds		500
	Total	7500

7.0 GLOSSARY AND METRIC CONVERSION GUIDE

A glossary at the end of the book defines terms which may not be familiar to some readers. These terms are printed in *italics* in the text.

All units in this book are metric and follow the International System of Units (SI). For the readers more familiar with the English and other common units, a metric conversion guide is found at the end of the book.

8.0 BACKGROUND READING

The following is a partial list of the most important references, periodicals, and conferences dealing with carbon.

8.1 General References

Chemistry and Physics of Carbon

Chemistry and Physics of Carbon, (P. L. Walker, Jr. and P. Thrower, eds.), Marcel Dekker, New York (1968)

Cotton, F. A. and Wilkinson, G., *Advanced Inorganic Chemistry*, Interscience Publishers, New York (1972)

Eggers, D. F., Gregory, N. W., Halsey, G. D., Jr. and Rabinovitch, B. S., *Physical Chemistry*, John Wiley & Sons, New York (1964)

Huheey, J. E., *Inorganic Chemistry*, Third Edition, Harper & Row, New York (1983)

Jenkins, G. M. and Kawamura, K., *Polymeric Carbons*, Cambridge University Press, Cambridge, UK (1976)

Mantell, C. L., *Carbon and Graphite Handbook,* Interscience, New York (1968)

Van Vlack, L. H., *Elements of Materials Science and Engineering,* 4th ed., Addison-Wesley Publishing Co., Reading MA (1980)

Wehr, M. R., Richards, J. A., Jr., and Adair, T. W., III, *Physics of the Atom,* Addison-Wesley Publishing Co., Reading, MA (1978)

Carbon Fibers

Donnet, J-B. and Bansal, R. C., *Carbon Fibers,* Marcel Dekker Inc., New York (1984)

Carbon Fibers Filaments and Composites (J. L. Figueiredo, et al., eds.), Kluwer Academic Publishers, The Netherlands (1989)

Dresselhaus, M. S., Dresselhaus, G., Sugihara, K., Spain, I. L., and Goldberg, H. A., *Graphite Fibers and Filaments,* Springer Verlag, Berlin (1988)

Diamond

Applications of Diamond Films and Related Materials (Y. Tzeng, et al., eds.), Elsevier Science Publishers, 623-633 (1991)

Davies, G., *Diamond,* Adams Hilger Ltd., Bristol UK (1984)

The Properties of Diamond (J. E. Field, ed.), 473-499, Academic Press, London (1979)

8.2 Periodicals

- Applied Physics Letters
- Carbon
- Ceramic Bulletin
- Ceramic Engineering and Science Proceedings
- Diamond and Related Materials (Japan)
- Diamond Thin Films (Elsevier)
- Japanese Journal of Applied Physics
- Journal of the American Ceramic Society

- Journal of the American Chemical Society
- Journal of Applied Physics
- Journal of Crystal Growth
- Journal of Materials Research
- Journal of Vacuum Science and Technology
- Materials Engineering
- Materials Research Society Bulletin
- Nature
- SAMPE Journal
- SAMPE Quarterly
- Science
- SPIE Publications
- Tanso (Tokyo)

8.3 Conferences

- Carbon Conference (biennial)
- International Conference on Chemical Vapor Deposition (CVD) of the Electrochemical Society (biennial)
- Composites and Advanced Ceramics Conference of the American Ceramic Society (annual)
- Materials Research Society Conference (annual)

REFERENCES

1.   Krauskopf, K. B., *Introduction to Geochemistry*, McGraw-Hill Book Co., New York (1967)
2.   *Chart of the Atoms*, Sargent-Welch Scientific Co., Skokie, IL (1982)
3.   Hare, J. P. and Kroto, H. W., A Postbuckminsterfullerene View of Carbon in the Galaxy, *Acc. Chem. Res.*, 25:106-112 (1992)
4.   Davies, G., *Diamond*, Adam Hilger Ltd., Bristol, UK (1984)
5.   *Data Bank*, G.A.M.I., Gorham, ME (1992)

2

The Element Carbon

1.0 THE STRUCTURE OF THE CARBON ATOM

1.1 Carbon Allotropes and Compounds

The primary objective of this book is the study of the element carbon itself and its polymorphs, i.e., graphite, diamond, fullerenes, and other less common forms. These allotropes (or polymorphs) have the same building block, the carbon atom, but their physical form, i.e., the way the building blocks are put together, is different. In other words, they have distinct molecular or crystalline forms.

The capability of an element to combine its atoms to form such allotropes is not unique to carbon. Other elements in the fourth column of the periodic table, silicon, germanium, and tin, also have that characteristic. However carbon is unique in the number and the variety of its allotropes.

The properties of the various carbon allotropes can vary widely. For instance, diamond is by far the hardest-known material, while graphite can be one of the softest. Diamond is transparent to the visible spectrum, while graphite is opaque; diamond is an electrical insulator while graphite is a conductor, and the fullerenes are different from either one. Yet these materials are made of the same carbon atoms; the disparity is the result of different arrangements of their atomic structure.

Just as carbon unites easily with itself to form polymorphs, it can also combine with hydrogen and other elements to give rise to an extraordinary number of compounds and *isomers* (i.e., compounds with the same composition but with different structures). The compounds of carbon and hydrogen and their derivatives form the extremely large and complex branch of chemistry known as organic chemistry. More than half-a-million organic compounds are identified and new ones are continuously discovered. In fact, far more carbon compounds exist than the compounds of all other elements put together.[1]

While organic chemistry is not a subject of this book, it cannot be overlooked since organic compounds play a major part in the processing of carbon polymorphs. Some examples of organic precursors are shown in Table 2.1.[2]

Table 2.1. Organic Precursors of Carbon Products

Precursors	Products
Methane	Pyrolytic graphite
Hydrocarbons Fluorocarbons Acetone, etc.	Diamond-like carbon Polycrystalline diamond
Rayon Polyacrylonitrile	Carbon fibers
Phenolics Furfuryl alcohol	Carbon-carbon Vitreous carbon
Petroleum fractions Coal tar pitch	Molded graphites Carbon fibers
Plants	Coal

In order to understand the formation of the allotropes of carbon from these precursors and the reasons for their behavior and properties, it is essential to have a clear picture of the atomic configuration of the carbon atom and the various ways in which it bonds to other carbon atoms. These are reviewed in this chapter.

1.2 The Structure of the Carbon Atom

All atoms have a positively charged nucleus composed of one or more protons, each with a positive electrical charge of +1, and neutrons which are electrically neutral. Each proton and neutron has a mass of one and together account for practically the entire mass of the atom. The nucleus is surrounded by electrons, moving around the nucleus, each with a negative electrical charge of -1. The number of electrons is the same as the number of protons so that the positive charge of the nucleus is balanced by the negative charge of the electrons and the atom is electrically neutral.

As determined by Schroedinger, the behavior of the electrons in their movement around the nucleus is governed by the specific rules of *standing waves*.[3] These rules state that, in any given atom, the electrons are found in a series of energy levels called *orbitals*, which are distributed around the nucleus. These orbitals are well defined and, in-between them, large ranges of intermediate energy levels are not available (or forbidden) to the electrons since the corresponding frequencies do not allow a standing wave.

In any orbital, no more than two electrons can be present and these must have opposite *spins* as stated in the *Pauli's exclusion principle*. A more detailed description of the general structure of the atom is given in Ref. 3, 4, and 5.

Nucleus and Electron Configuration of the Carbon Atom. The element carbon has the symbol C and an *atomic number* (or Z number) of 6, i.e., the neutral atom has six protons in the nucleus and correspondingly six electrons. In addition, the nucleus includes six neutrons (for the carbon-12 isotope, as reviewed in Sec. 2.0 below). The electron configuration, that is, the arrangement of the electrons in each orbital, is described as: $1s^2 2s^2 2p^2$. This configuration is compared to that of neighboring atoms in Table 2.2.

The notation $1s^2$ refers to the three *quantum numbers* necessary to define an orbital, the number "1" referring to the K or first shell (principal quantum number). The letter "s" refers to the sub-shell s (*angular momen-*

tum quantum number) and the superscript numeral "2" refers to the number of atoms in that sub-shell. There is only one orbital (the s orbital) in the K shell which can never have more than two electrons. These two electrons, which have opposite spin, are the closest to the nucleus and have the lowest possible energy. The filled K shell is completely stable and its two electrons do not take part in any bonding.

Table 2.2. Electron Configuration of Carbon and Other Atoms

Element		K	L		M			First Ionization
Symbol	Z	1s	2s	2p	3s	3p	3d	Potential (eV)
H	**1**	**1**						**13.60**
He	2	2						24.59
Li	3	2	1					5.39
Be	4	2	2					9.32
B	5	2	2	1				8.30
C	6	2	2	2				11.26
N	**7**	**2**	**2**	**3**				**14.53**
O	**8**	**2**	**2**	**4**				**13.62**
F	9	2	2	5				17.42
Ne	10	2	2	6				21.56
Na	11	2	2	6	1			5.14
Etc.								

Note: The elements shown in bold (H, N and O) are those which combine with carbon to form most organic compounds.

The next two terms, $2s^2$ and $2p^2$, refer to the four electrons in the L shell. The L shell, when filled, can never have more than eight electrons. The element neon has a filled L shell. The L-shell electrons belong to two different subshells, the s and the p, and the 2s and the 2p electrons have different energy levels (the number "2" referring to the L or second shell, and the letters "s" and "p" to the orbitals or *sub-shells*). The two 2s electrons have opposite spin and the two 2p electrons parallel spin. This view of the carbon atom is represented schematically in Fig. 2.1.

The configuration of the carbon atom described above refers to the configuration in its ground state, that is, the state where its electrons are in their minimum orbits, as close to the nucleus as they can be, with their lowest energy level.

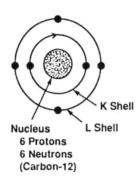

Nucleus
6 Protons
6 Neutrons
(Carbon-12)

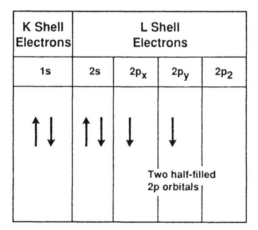

K Shell Electrons	L Shell Electrons		
1s	2s	$2p_x$	$2p_y$	$2p_2$

Note: Arrow indicates direction of electron spin

Figure 2.1. Schematic of the electronic structure of the carbon atom in the ground state.

Valence Electrons and Ionization Potential. In any given atom, the electrons located in the outer orbital are the only ones available for bonding to other atoms. These electrons are called the valence electrons. In the case of the carbon atom, the valence electrons are the two 2p orbitals. Carbon in this state would then be divalent, since only these two electrons are available for bonding.

Divalent carbon does indeed exist and is found in some highly reactive transient-organic intermediates such as the *carbenes* (for instance methylene). However, the carbon allotropes and the stable carbon compounds are not divalent but tetravalent, which means that four valence electrons are present.[6] How this increase in valence electrons occurs is reviewed in Sec. 3.0.

The carbon valence electrons are relatively easily removed from the carbon atom. This occurs when an electric potential is applied which accelerates the valence electron to a level of kinetic energy (and corresponding momentum) which is enough to offset the binding energy of this electron to the atom. When this happens, the carbon atom becomes ionized forming a positive ion (cation). The measure of this binding energy is the *ionization potential*, the first ionization potential being the energy necessary to remove the first outer electron, the second ionization potential, the second outer second electron, etc. The ionization energy is the product of the elementary charge (expressed in volts) and the ionization potential, expressed in *electron volts*, eV (one eV being the unit of energy accumulated by a particle with one unit of electrical charge while passing though a potential difference of one volt).

The first ionization potentials of carbon and other atoms close to carbon in the Periodic Table are listed in Table 2.2. It should be noted that the ionization energy gradually (but not evenly) increases going from the first element of a given shell to the last. For instance, the value for lithium is 5.39 V and for neon, 21.56 V. It is difficult to ionize an atom with a complete shell such as neon, but easy to ionize one with a single-electron shell such as lithium.

As shown in Table 2.2 above, carbon is located half-way between the two noble gases, helium and neon. When forming a compound, carbon can either lose electrons and move toward the helium configuration (which it does when reacting with oxygen to form CO_2), or it can gain electrons and move toward the neon configuration (which it does when combining with other carbon atoms to form diamond).

The six ionization potentials of the carbon atom are shown in Table 2.3.

Table 2.3. Ionization Potentials of the Carbon Atom

Number	Shell	Orbital	Potential, V
1st	L	2p	11.260
2d	L	2p	24.383
3d	L	2s	47.887
4th	L	2s	64.492
5th	K	1s	392.077
6th	K	1s	489.981

As shown in Table 2.3, in an element having a low atomic number such as carbon, the difference in energy of the electrons within one shell, in this case between the 2s and 2p electrons, is relatively small compared to the differences in energy between the electrons in the various shells, that is between the K shell ($1s^2$ electrons) and the L shell ($2s^2$ and $2p^2$ electrons). As can be seen, to remove the two electrons of the K shell requires considerably more energy than to remove the other four electrons.

1.3 Properties and Characteristics of the Carbon Atom

The properties and characteristics of the carbon atom are summarized in Table 2.4.

Table 2.4. Properties and Characteristics of the Carbon Atom

- Z (atomic number = number of protons or electrons): 6
- N (number of neutrons): 6 or 7 (common isotopes)
- A (Z + N or number of nucleons or mass number): 12 or 13
- Atomic Mass: 12.01115 amu (see below)
- Atomic Radius: 0.077 nm (graphite structure) (see below)
- First Ionization Potential: v = 11.260
- Quantum Number of Last Added Electron: n = 2, l = 1
- Outermost Occupied Shell: L

Atomic Mass (Atomic Weight): The element carbon is used as the basis for determining the atomic mass unit. The atomic mass unit (amu) is, by definition, 1/12th of the atomic mass of the carbon-12 (^{12}C) isotope. This definition was adopted in 1961 by International Union of Pure and Applied Chemistry. The atomic mass unit is, of course, extremely small compared to the standard concept of mass: it takes 0.6022×10^{24} amu to make one gram (this number is known as Avogadro's number or N). As will be shown in Sec. 2.0 below, natural carbon contains approximately 98.89% ^{12}C and 1.11% of the heavier ^{13}C. As a result, the atomic mass of the average carbon atom is 12.01115 amu (see Sec. 2.0).

Atomic Radius: The atomic radius of carbon is half the equilibrium distance between two carbon atoms of the planar graphite structure. Carbon has one of the smallest radii of all the elements as shown in Table 2.5. All elements not shown in this table have larger radii.

Table 2.5. Atomic Radii of Selected Elements

Element	Atomic Radius nm
Hydrogen	0.046
Helium	0.176
Lithium	0.152
Beryllium	0.114
Boron	0.046
Carbon	**0.077**
Nitrogen	0.071
Oxygen	0.060
Fluorine	0.06

2.0 THE ISOTOPES OF CARBON

2.1 Characteristics of the Carbon Isotopes

The isotopes of an element have the same atomic number Z, i.e., the same number of protons and electrons and the same electron configuration.

However they have a different number of neutrons, and therefore a different mass number; the mass number (or atomic weight) is the sum of the protons and neutrons, represented by the symbol "A".

The element carbon has seven isotopes, which are listed in Table 2.6. The most common isotope by far is ^{12}C which has six neutrons. The others have from four to ten neutrons (^{10}C to ^{16}C).

Table 2.6. Properties of the Carbon Isotopes[7]

Isotope	% Natural Abundance	Atomic Mass	Half-life	Decay Modes	Decay Energy (MeV)	Particle Energies (MeV)
^{10}C	-	-	19.45 s	β^+	3.61	1.87
^{11}C	-	-	20.3 m	β^+,EC	1.98	0.98
^{12}C	98.89	12.000	stable	-	-	-
^{13}C	1.108	13.00335	stable	-	-	-
^{14}C	-	-	5730 yr.	β^-	0.156	0.156
^{15}C	-	-	2.4 s	β^-	9.8	9.82
						4.51
^{16}C	-	-	0.74 s	β^-,n		

Note: β^- = negative beta emission
β^+ = positive beta emission
EC = orbital electron capture
n   = neutron emission

Carbon-12 and carbon-13 are stable isotopes, that is, they do not spontaneously change their structure and disintegrate. The other five carbon isotopes are radioactive, i.e., they decay spontaneously by the emission of β particles, which are either an electron (β^-) or a *positron* (β^+) and are generated from the splitting of a neutron. The average rate of disintegration is fixed, regardless of any changes that may occur in the chemical or physical conditions of the atom. Disintegration of a radioactive isotope is measured in terms of half-life, which is the time required for the original number of radioactive isotopes to be reduced to one-half.

As shown in Table 2.4, the ^{10}C, ^{11}C, ^{15}C and ^{16}C isotopes have short half-lives, and their practical use is therefore limited. On the other hand, ^{14}C has a long half-life and is a useful isotope with important applications (see below).

The atomic structures of ^{12}C, ^{13}C and ^{14}C are shown schematically in Fig. 2.2.

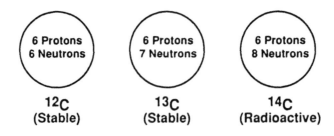

6 Protons 6 Neutrons	6 Protons 7 Neutrons	6 Protons 8 Neutrons
^{12}C (Stable)	^{13}C (Stable)	^{14}C (Radioactive)

Figure 2.2. Major carbon isotopes. Note equal number of protons but different number of neutrons.[8]

2.2   Carbon Dating with Carbon-14

The radioactive decay of ^{14}C and of other radioactive isotopes, such as uranium-235 and -238, thorium-232, rubidium-87 and potassium-K40, provide a reliable way of dating materials. Carbon-14 can only be used to date carbonaceous compounds. Its long half-life of 5730 years permits accurate dating for up to 30,000 years. This period is approximately equal to five half-lives, after which only 1/32nd of the original amount of ^{14}C remains, which is no longer sufficient to permit precise measurements.[8][9]

Mechanism of Formation and Decay of ^{14}C. The chemist Willard F. Libby discovered in 1946 that ^{14}C is continuously being formed in the earth's atmosphere by the reaction of the major nitrogen isotope, ^{14}N, with highly energetic neutrons originating as a secondary radiation from cosmic rays.[9] In this reaction the ^{14}N atom gains a neutron (going from seven to eight) and loses a proton (going from seven to six), thus decreasing in atomic number and becoming ^{14}C.

As mentioned above, ^{14}C is a radioactive isotope and decays spontaneously by emitting β⁻ particles, thus forming a nitrogen atom, as shown schematically in Fig. 2.3. The processes of formation and decay are in equilibrium in the atmosphere and the amount of ^{14}C remains essentially constant at a low level. Much of this ^{14}C is found in the atmospheric carbon dioxide.

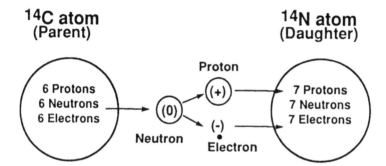

^{14}C atom
(Parent)

^{14}N atom
(Daughter)

Proton

6 Protons
6 Neutrons
6 Electrons

(0)

(+)

(-)

Neutron

Electron

7 Protons
7 Neutrons
7 Electrons

A neutron of the ^{14}C atom spontaneously forms a proton and a beta particle which becomes the new electron of the nitrogen atom.

Figure 2.3. Decay of the ^{14}C radioactive carbon isotope.[8]

Carbon-14 in Living and Dead Matter. Plants continuously absorb CO_2 and, consequently, maintain a constant level of ^{14}C in their tissues. Animals consume plants (or other plant-eating animals) and thus every living thing contains carbon that includes a small amount of ^{14}C in essentially the same ratio as the ^{14}C in the atmosphere. This amount is only 1×10^{-8} of the amount of ^{12}C.

After death, ^{14}C is no longer replaced and, through the radioactive decay process, remains in dead matter in a steadily diminishing amount as time goes by. This amount can be measured (and the years since death readily computed) by counting the number of β⁻ particles emitted by the remaining ^{14}C atoms (called the ^{14}C activity) and comparing it to the activity of a contemporary living sample.

Applications of Carbon Dating. Dating with the ^{14}C isotope is a practical and widely used method of dating carbonaceous materials (Fig. 2.4). It is used extensively in archeology, paleontology, and other disciplines, to date wood from Egyptian and Etruscan tombs or determine the age of the Dead-Sea scrolls and of prehistoric animals and plants, to mention only some well-known examples. By the dating of trees caught in advancing glaciers, it has been possible to calculate the glacial cycles of the earth in the last 30,000 years.

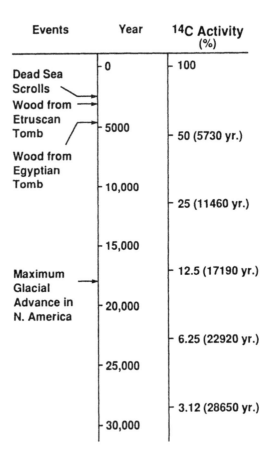

Figure 2.4. Dating sequence of the ^{14}C carbon isotope.

2.3   The ^{12}C and ^{13}C Isotopes

There is good experimental evidence that the properties of carbon allotropes or compounds are affected by the isotopic composition of the carbon atoms, as shown by the following examples.

One process for synthesizing diamond uses methane in which the carbon atom is a carbon-12 isotope enriched to 99.97% ^{12}C. In this process, diamond is deposited by chemical vapor deposition (CVD) in a microwave plasma (see Ch. 13 for a description of the process). The resulting ^{12}C diamond is reported to have a thermal conductivity 50% higher than that of natural diamond which has the normal ratio of ^{12}C and ^{13}C of about 100/1.[10]

The other example is related to the natural process of photosynthesis in organic matter. Photosynthesis is isotope-selective, i.e., less ^{13}C is absorbed proportionally so that the carbon from organic sources is slightly poorer in ^{13}C than inorganic carbon (1.108 % vs. 1.110 % ^{13}C).

This selective absorption is important in geochemical studies since measuring the amount of ^{13}C provides a good evidence of the origin of the carbon. For instance, the mineral calcite, which is a limestone composed mostly of calcium carbonate ($CaCO_3$) found in the cap rock of salt domes, is low in ^{13}C compared to most other limestones which have an inorganic origin. This indicates that the carbon in the calcite came from petroleum (an organic source), rather than from non-organic sources.

Most limestones are formed from the bicarbonate ion of sea water, HCO_3^-, (which in turn comes from atmospheric CO_2) and have the normal ^{13}C content. The salt-dome calcite, on the other hand, is formed by the combination of a calcium ion, Ca^{++}, and the CO^2 resulting from the oxidation of petroleum (hence from an organic source with less ^{13}C.[8][11] These formation processes are shown schematically in Fig. 2.5.

3.0   HYBRIDIZATION AND THE sp^3 CARBON BOND

3.1   The Carbon Bond

The characteristics and properties of the single carbon atom were described in the preceding sections. This section is a review of the ways carbon atoms bond together to form solids, such as diamond, graphite, and other carbon polymorphs.

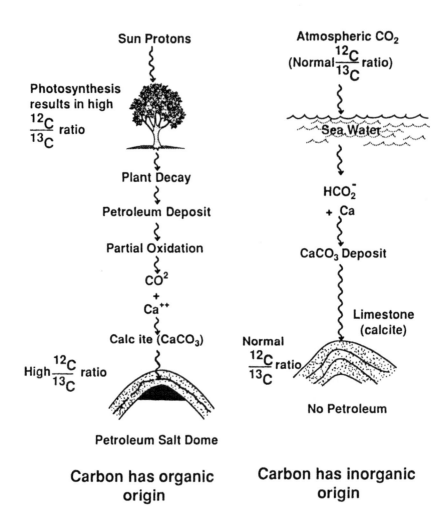

Figure 2.5. Organic origin of carbon in petroleum salt domes.

A chemical bond is formed when an electron becomes sufficiently close to two positive nuclei to be attracted by both simultaneously (unless the attraction is offset by repulsion from other atoms within the molecule). In the case of carbon molecules, this bonding is covalent (that is, neighboring atoms share electrons) and can take several forms: the sp^3, sp^2 and sp orbital bonds.

3.2 Hybridization of the Carbon Atom

Electron Orbitals. As mentioned previously, the electron configuration of the six electrons of the carbon atom in the ground state (i.e., a single atom) is $1s^2 2s^2 2p^2$, that is, two electrons are in the K shell (1s) and four in the L shell, i.e., two in the 2s orbital and two in the 2p orbital (Fig. 2.1).

It should be stressed at this stage that no electron in an atom or a molecule can be accurately located. The electron *wave function* Ψ establishes the probability of an electron being located in a given volume with the nucleus being the origin of the coordinate system. Mathematically speaking, this function has a finite value anywhere in space, but the value of the function becomes negligible at a distance of a few angstroms from the nucleus. For all practical purposes, the volume where the electron has the highest probability of being located is well defined and is usually represented as a small shaded volume.[1] What is uncertain is the precise location within this volume.

Ground-State Carbon Orbitals. The carbon-atom orbitals in the ground state can be visualized as shown graphically Fig. 2.6. The wave-function calculations represent the s orbital as a sphere with a blurred or fuzzy edge that is characteristic of all orbital representation. As a sphere, the s orbital is non-directional. The 2p orbital can be represented as an elongated barbell which is symmetrical about its axis and directional.

The Carbon Hybrid sp^3 Orbital. The $1s^2 2s^2 2p^2$ configuration of the carbon atom does not account for the tetrahedral symmetry found in structures such as diamond or methane (CH_4) where the carbon atom is bonded to four other carbon atoms in the case of diamond, or to four atoms of hydrogen in the case of methane. In both cases, the four bonds are of equal strength.

In order to have a electron configuration that would account for this symmetry, the structure of the carbon atom must be altered to a state with four valence electrons instead of two, each in a separate orbital and each

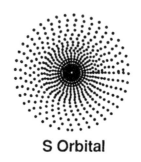

S Orbital

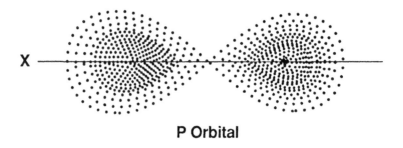

P Orbital

Figure 2.6. Schematic representation of the "s" and "p" orbitals.

with its spin uncoupled from the other electrons. This alteration occurs as a result of the formation of *hybrid atomic orbitals*, in which the arrangement of the electrons of the L shell of the atom in the ground state is modified as one of the 2s electron is promoted (or lifted) to the higher orbital 2p as shown in Fig. 2.7. These new orbitals are called hybrids since they combine the 2s and the 2p orbitals. They are labeled sp^3 since they are formed from one s orbital and three p orbitals.

In this hybrid sp^3 state, the carbon atom has four $2sp^3$ orbitals, instead of two 2s and two 2p of the ground-state atom and the valence state is raised from two to four. The calculated sp^3 electron-density contour is shown in Fig. 2.8 and a graphic visualization of the orbital, in the shape of an electron cloud, is shown in Fig. 2.9.[12] This orbital is asymmetric, with most of it concentrated on one side and with a small tail on the opposite side.

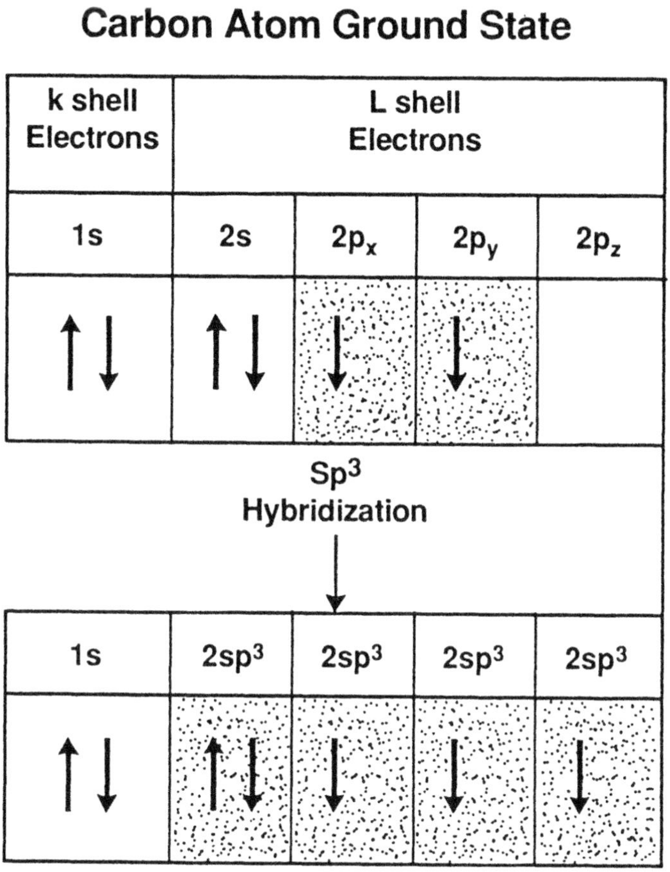

Carbon Atom Ground State

k shell Electrons	L shell Electrons			
1s	2s	$2p_x$	$2p_y$	$2p_z$
↑↓	↑↓	↓	↓	

Sp³
Hybridization
↓

1s	2sp³	2sp³	2sp³	2sp³
↑↓	↑↓	↓	↓	↓

Figure 2.7. The sp³ hybridization of carbon orbitals. Shaded electrons are valence electrons (divalent for ground state, tetravalent for hybrid state). Arrow indicates direction of electron spin.

As shown in Figs. 2.8 and 2.9 (and in following related figures), the lobes are labeled either + or -. These signs refer to the sign of the wave function and not to any positive or negative charges since an electron is always negatively charged. When an orbital is separated by a *node*, the signs are opposite.

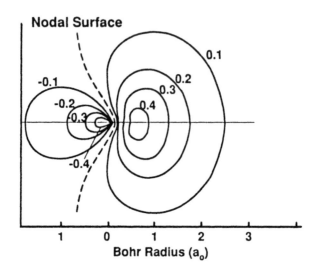

Figure 2.8. Electron density contour of sp^3 orbital.[12]

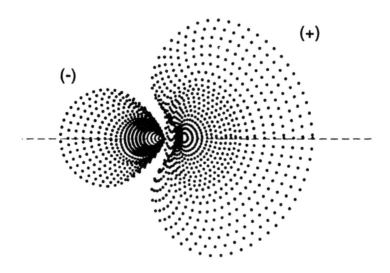

Figure 2.9. Cloud representation of sp^3 hybrid orbital.[12]

A graphic visualization of the formation of the sp^3 hybridization is shown in Fig. 2.10. The four hybrid sp^3 orbitals (known as tetragonal hybrids) have identical shape but different spatial orientation. Connecting the end points of these vectors (orientation of maximum probability) forms a regular tetrahedron (i.e., a solid with four plane faces) with equal angles to each other of 109° 28'.

The energy required to accomplish the sp^3 hybridization and raise the carbon atom from the ground state to the corresponding valence state V_4 is 230 kJ mol^{-1}. This hybridization is possible only because the required energy is more than compensated by the energy decrease associated with forming bonds with other atoms.

The hybridized atom is now ready to form a set of bonds with other carbon atoms. It should be stressed that these hybrid orbitals (and indeed all hybrid orbitals) are formed only in the bonding process with other atoms and are not representative of an actual structure of a free carbon atom.[13]

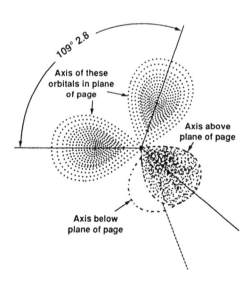

Figure 2.10. Tetrahedral hybridization axes of the four sp^3 orbitals. Negative lobes omitted for clarity.

3.3   The Carbon Covalent sp³ Bond

As mentioned above, carbon bonding is covalent and in the case of the sp³ bonding, the atoms share a pair of electrons. The four sp³ valence electrons of the hybrid carbon atom, together with the small size of the atom, result in strong covalent bonds, since four of the six electrons of the carbon atom form bonds.

The heavily lopsided configuration of the sp³ orbital allows a substantial overlap and a strong bond when the atom combines with a sp³ orbital from another carbon atom since the concentration of these bonding electrons between the nuclei minimizes the nuclear repulsion and maximizes the attractive forces between themselves and both nuclei. This bond formation is illustrated in Fig. 2.11.   By convention, a directional (or *stereospecific*) orbital such as the sp³ is called a sigma (σ) orbital, and the bond a sigma bond.

Each tetrahedron of the hybridized carbon atom (shown in Fig. 2.10) combines with four other hybridized atoms to form a three-dimensional, entirely covalent, lattice structure, shown schematically in Fig. 2.12. From the geometrical standpoint, the carbon nucleus can be considered as the center of a cube with each of the four orbitals pointing to four alternating corners of the cube.   This structure is the basis of the diamond crystal (see Ch. 11).

A similar tetrahedral bonding arrangement is also found in the methane molecule where the hybridized carbon atom is bonded to four hydrogen atoms. Four molecular orbitals are formed by combining each of the carbon sp³ orbitals with the orbital of the attached hydrogen atom (Fig. 2.13). The carbon tetrachloride molecule (CCl₄) is similar.

The tetragonal angle of 109° 28′ of the sigma-bond molecules must be considered as a time-averaged value since it changes continuously as the result of thermal vibrations.   The sigma-bond energy and the bond length will vary depending on the kind of atom which is attached to the carbon atom.   Table 2.7 shows the bond energy and the bond length of various carbon couples. The bond energy is the energy required to break one mole of bonds.   An identical amount of energy is released when the bond is formed. Included are the double and triple carbon bonds and other carbon bonds which will be considered later.

The sp³ bonds listed in Table 2.7 are found in all *aliphatic compounds* which are organic compounds with an open-ended chain structure and include the paraffin, olefin and acetylene hydrocarbons, and their derivatives.

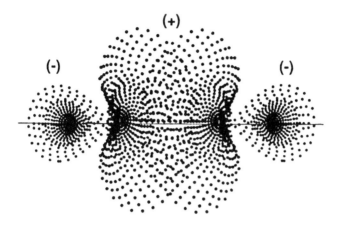

Figure 2.11. The sp^3 hybrid orbital bonding (sigma bond) showing covalent bonding.

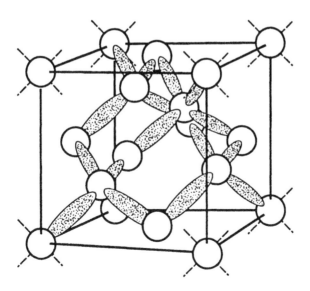

Figure 2.12. Three-dimensional representation of sp^3 covalent bonding (diamond structure). Shaded regions are regions of high electron probabilities where covalent bonding occurs.

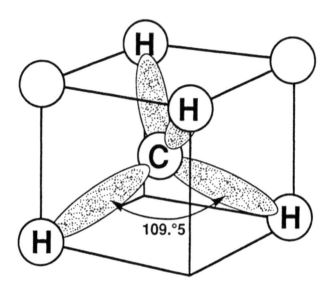

Figure 2.13. Three-dimensional representation of the methane molecule (CH₄) with sigma (sp³) bonding. Shaded regions are regions of high electron probabilities where covalentbonding occurs.[4]

Table 2.7. Carbon-Couples Bond Energies and Lengths

Bond	Hybrid type	Approximate bond energy* kJ/mole	kcal/mole	Bond length nm
C-C	sp³	370	88	0.154
C=C	sp²	680	162	0.13
C≡C	sp	890	213	0.12
C-H	sp³	435	104	0.109
C-Cl	sp³	340	81	0.18
C-N	sp³	305	73	0.15
C-O	sp³	360	86	0.14

* Energy required to break one mole of bonds (Avogadro's number)

4.0   THE TRIGONAL sp^2 AND DIGONAL sp CARBON BONDS

4.1   The Trigonal sp^2 Orbital

In addition to the sp^3-tetragonal hybrid orbital reviewed in Sec. 3 above, two other orbitals complete the series of electronic building blocks of all carbon allotropes and compounds: the sp^2 and the sp orbitals.

Whereas the sp^3 orbital is the key to diamond and aliphatic compounds, the sp^2 (or trigonal) orbital is the basis of all graphitic structures and *aromatic compounds*.

The mechanism of the sp^2 hybridization is somewhat different from that of the sp^3 hybridization. The arrangement of the electrons of the L shell of the atom in the ground state is modified as one of the 2s electrons is promoted and combined with two of the 2p orbitals (hence the designation sp^2), to form three sp^2 orbitals and an unhybridized free (or delocalized) p orbital electron as shown in Fig. 2.14. The valence state is now four (V4).

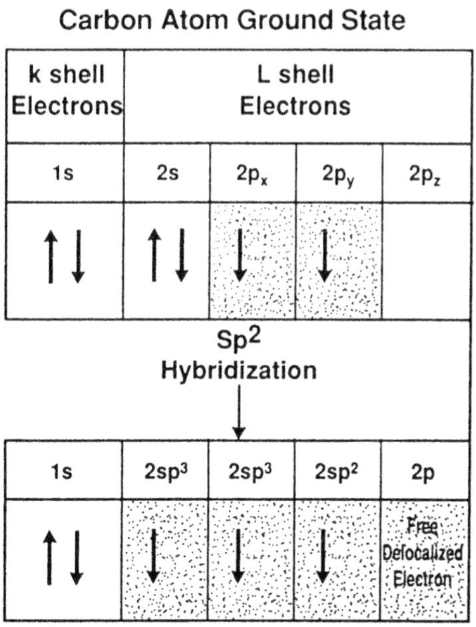

Carbon Atom Ground State

Figure 2.14. The sp^2 hybridization of carbon orbitals. Shaded electrons are valence electrons (divalent for ground state and tetravalent for hybrid state).

The calculated electron-density contour of the sp^2 orbital is similar in shape to that of the sp^3 orbital shown in Figs. 2.8 and 2.9. These three identical sp^2 orbitals are in the same plane and their orientation of maximum probability forms a 120° angle from each other as shown in Fig. 2.15.

The fourth orbital, i.e., the delocalized non-hybridized p electron, is directed perpendicularly to the plane of the three sp^2 orbitals and becomes available to form the subsidiary pi (π) bond with other atoms.

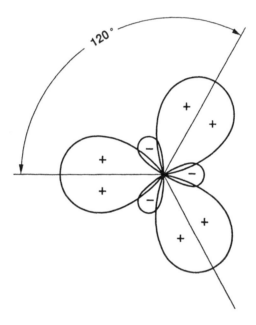

Figure 2.15. Planar section of the sp^2 hybrid orbitals of the carbon atom.

4.2   The Carbon Covalent sp^2 Bond

Like the sp^3 bond, the sp^2 bond is covalent. It is a strong bond, because of the three sp^2 valence electrons and the small size of the atom.

The lopsided configuration of the sp^2 orbital allows a substantial overlap with other sp^2 orbitals. This overlap is similar to the sp^3 overlap illustrated in Fig. 2.11, except that it is more pronounced, with a shorter bond length and higher bond energy, as shown in Table 2.7. Like the sp^3 orbital, the sp^2 is directional and is called a sigma (σ) orbital, and the bond a sigma bond.

Each sp²-hybridized carbon atom combines with three other sp²-hybridized atoms to form a series of hexagonal structures, all located in parallel planes as shown schematically in Fig. 2.16. The fourth valency, that is, the free delocalized electron, is oriented perpendicular to this plane as illustrated in Fig. 2.17. Unlike the sigma (σ) orbital, it is non-symmetrical and is called by convention a pi (π) orbital. It is available to form a subsidiary pi (π) bond.

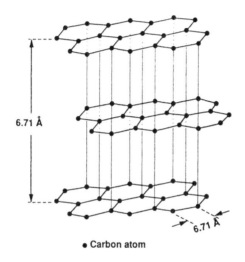

• Carbon atom

Figure 2.16. Three-dimensional schematic of the graphite structure.

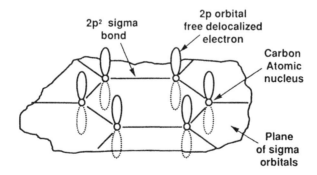

Figure 2.17. Schematic of the sp² hybridized structure of graphite showing the sigma bonds and the 2p freeelectrons (above and below the sigma orbitals plane).

In an sp² structure such as graphite, the delocalized electrons can move readily from one side of the plane layer to the other but cannot easily move from one layer to another. As a result, graphite is anisotropic. The sp²-hybridized structure of graphite will be reviewed in more detail in Ch. 3, Sec. 1.2.

4.3   The Digonal-sp Orbital and the sp Bond

The sp orbital (known as a digonal orbital) is a merger of an s and a p orbital which consists of two lobes, one large and one small, as illustrated in Fig. 2.18. An sp bond consists of two sp orbitals which, because of mutual repulsion, form an angle of 180° and, consequently, the sp molecule is linear. The bond, like all overlap bonds, is a sigma (σ) bond and has high strength. The sp orbitals account for two of the electrons of the carbon atom. The other two valence electrons are free, delocalized pi (π) orbital electrons which are available to form subsidiary pi (π) bonds in a manner similar to the sp² hybridization.

Examples of molecules having sp bonds are the gas acetylene, HC≡CH, and the carbynes, (C≡C)ₙ, which are cross-linked linear-chain carbon polytypes, usually unstable.[14]

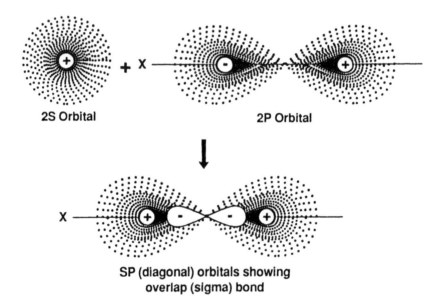

2S Orbital   2P Orbital

SP (diagonal) orbitals showing
overlap (sigma) bond

Figure 2.18. Formation of the sp hybrid orbital and sp sigma bond.

4.4   The Carbon-Hydrogen Bond

The carbon-hydrogen bond plays as important part in the mechanism of pyrolysis of carbon compounds and in the formation of graphite and diamond (the pyrolysis process is reviewed in Ch. 4).

The energy and length of the carbon-hydrogen bond are related to the type of hybridization of the carbon atom. The hybridization can be sp^3, sp^2 or sp as shown in Table 2.8.[5]

Table 2.8.   Properties of the Carbon-Hydrogen Bond and Effect of Hybridization

Molecule	Bond type	Approximate bond energy* (kJ/mole)	Hybrid bond length (nm)
CH radical	p	347	0.1120
CH_4 (methane)	sp^3	434	0.1094
C_2H_4 (ethylene)	sp^2	442	0.1079
C_2H_2 (acetylene)	sp	506	0.1057

* Energy required to break one mole of bonds (Avogadro's number)

5.0   CARBON VAPOR MOLECULES

At high temperature, carbon vaporizes to form a gas. This gas is a mixture of single carbon atoms and *diatomic* and *polyatomic molecules*, that is, molecules containing two, three, four, or more carbon atoms. These gaseous constituents are usually designated as C_1, C_2, C_3, etc.

The understanding of the composition and behavior of these carbon vapors, the accurate measurement of their heat of formation, and the precise determination of their ratio are essential to calculate the heat of formation of organic compounds, i.e., the energies of all bonds involving carbon.

The vaporization of carbon is a major factor in the *ablation* of carbon. This ablation is the basic phenomena that controls the performance of rocket-nozzle throats, reentry nose cones and other components exposed to extremely high temperatures (see Ch. 9). The rate of ablation is related to the composition of the carbon vapor formed during ablation and to the heat of formation of the various carbon-vapor species and their evaporation coefficient.

Recent and accurate mass-spectrographic measurements of the energy required to vaporize graphite to the monoatomic gas C_1 give a value of 710.51 kJ mol^{-1} (171.51 kcal mol^{-1}).[15] Values for the heat of formation of the molecular vapor species of carbon shown in Table 2.9.

Table 2.9. Heat of Formation of Carbon Molecules[15]

Molecule	Heat of Formation (kJ/mole)
C_2	823
C_3	786
C_4	1005
C_5	1004
C_6	1199
C_7	1199
C_8	1417
C_9	1396
C_{10}	1642

C_1, C_2, and particularly C_3 are the dominant species in the equilibrium vapor in the temperature range of 2400 - 2700 K as shown in the Arrhenius plot of the partial pressure of these species in Fig. 2.19.[15][16] The contribution of C_6 and larger molecules to the vapor pressure is small and generally of no practical import. The general structure of these carbon molecule is believed to consist of double carbon bonds, :C=C::::C=C: (the so-called cumulene structure) which have delocalized bondings and an axial symmetry. Larger molecules, i.e., the fullerenes, are reviewed below.

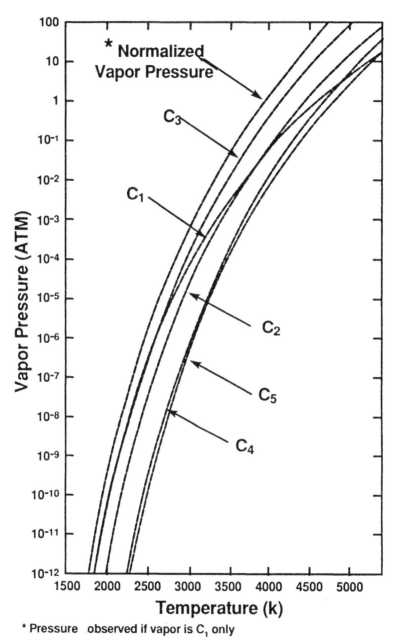

Figure 2.19. Vapor pressure of carbon species.

6.0   THE CARBON ALLOTROPES

6.1   The Carbon Phase Diagram

The carbon phase diagram is shown in Fig. 2.20.[6] Another expression of the T-P phase diagram, showing the calculated total vapor pressure of carbon, is shown in Fig. 3.7 of Ch. 3.

Carbon vaporizes at 4800 K at a pressure of 1000 atmospheres, which is the area where diamond is stable. The high-pressure conversion of diamond from graphite occurs at temperatures of approximately 3000 K and pressures above 125 kbars (in the absence of catalyst) and will be reviewed in Ch. 12.

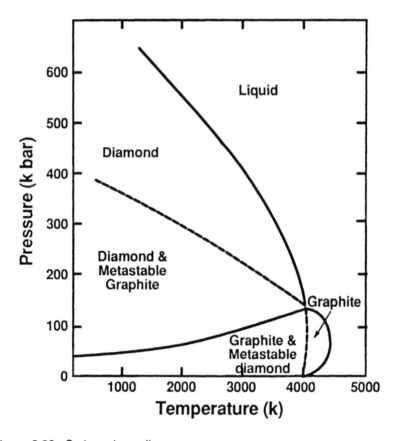

Figure 2.20. Carbon phase diagram.

6.2 Allotropic Forms

In the preceding sections, the various ways that carbon atoms bond together to form solids were reviewed. These solids are the allotropes (or polymorphs) of carbon, that is, they have the same building block—the element carbon—but with different atomic hybrid configurations: sp^3 (tetragonal), sp^2 (trigonal) or sp (digonal).

These allotropic solids can be classified into three major categories: (i) the sp^2 structures which include graphite, the graphitic materials, amorphous carbon, and other carbon materials (all reviewed in Ch. 3), (ii) the sp^3 structures which include diamond and lonsdaleite (a form detected in meteorites), reviewed in Ch. 11, and (iii) the Fullerenes (see Ch. 15).

These allotropes are sometimes found in combination such as some diamond-like carbon (DLC) materials produced by low-pressure synthesis, which are actually mixtures of microcrystalline diamond and graphite (see Ch. 14).

Recent investigations have revealed the existence of a series of diamond polytypes such as the 6-H hexagonal diamond. The structure and properties of these polytypes are reviewed in Ch. 11.[17][18] Also under investigation is a hypothetical phase of carbon based on a three-dimensional network but with sp^2 bonds. This phase could be harder than diamond, at least in theory.[19] A carbon phase diagram incorporating these new polytypes has yet to be devised.

6.3 The Fullerene Carbon Molecules

The recently discovered family of fullerene carbon molecules are considered another major allotropic form of carbon that combines both sp^2 and sp^3 bonds. These molecules are still in the early stages of investigation and it will be some time before practical applications are found. The fullerenes are reviewed in Ch. 15.

REFERENCES

1.   Cram, D. J. and Hammond, G. S., *Organic Chemistry*, McGraw-Hill Book Co., New York (1964)

2.   Jenkins, G. M. and Kawamura, K., *Polymeric Carbons*, Cambridge University Press, Cambridge, UK (1976)

3.   Wehr, M. R., Richards, J. A., Jr. and Adair, T. W., III, *Physics of the Atom*, Addison-Wesley Publishing Co., Reading, MA (1978)

4.   Van Vlack, L. H., *Elements of Materials Science and Engineering*, 4th ed., Addison-Wesley Publishing Co., Reading, MA (1980)

5.   Eggers, D. F., Gregory, N. W., Halsey, G. D., Jr. and Rabinovitch, B. S., *Physical Chemistry*, John Wiley & Sons, New York (1964)

6.   Cotton, F. A. and Wilkinson, G., *Advanced Inorganic Chemistry*, Interscience Publishers, New York (1972)

7.   *Handbook of Chemistry and Physics*, 65th ed., CRC Press, Boca Raton, FL (1985)

8.   Press, F. and Siever R., *Earth*, W.H. Freeman & Co., San Francisco (1974)

9.   Asimov, A., *Understanding Physics*, Vol. 3, Dorset Press (1988)

10. *Ceramic Bull.*, 69(10):1639 (1990)

11. Krauskopf, K. B., *Introduction to Geochemistry*, McGraw-Hill Book Co., New York (1967)

12. Huheey, J. E., *Inorganic Chemistry*, 3rd. ed., Harper & Row, New York (1983)

13. March, J., *Advanced Inorganic Chemistry*, John Wiley & Sons, New York (1985)

14. Korshak, V. V., et al, *Carbon*, 25(6):735-738 (1987)

15. Palmer, H. B. and Shelef, M., *Chemistry and Physics of Carbon*, (P. L. Walker, Jr., ed.), Vol.4, Marcel Dekker, New York (1968)

16. Mantell, C. L., *Carbon and Graphite Handbook*, Interscience, New York (1968)

17. Spear, K. E., Phelps, A. W., and White, W. B., *J. Mater. Res.*, 5(11):2271-85 (Nov. 1990)

18. Bundy, F. P. and Kasper, J. S., *J. of Chem. Physics*, 46(9):3437-3446 (1967)

19. Tamor, M. A. and Hass, K. C., *J. Mater. Res.*, Vol. 5(11):2273-6 (Nov. 1990)

3

Graphite Structure and Properties

1.0 THE STRUCTURE OF GRAPHITE

1.1 General Considerations and Terminology

The origin of the word "graphite" is the Greek word "graphein" which means "to write". Indeed, graphite has been used to write (and draw) since the dawn of history and the first pencils were manufactured in England in the 15th century. In the 18th century, it was demonstrated that graphite actually is an allotrope of carbon.

Graphite is remarkable for the large variety of materials that can be produced from its basic form such as extremely strong fibers, easily sheared lubricants, gas-tight barriers, and gas adsorbers. All these diverse materials have one characteristic in common: they are all built upon the trigonal sp^2 bonding of carbon atoms.

Strictly speaking, the term "graphite" by itself describes an ideal material with a perfect graphite structure and no defects whatsoever. However, it is also used commonly, albeit incorrectly, to describe graphitic materials. These materials are either "graphitic carbons", that is, materials consisting of carbon with the graphite structure, but with a number of structural defects, or "non-graphitic carbons", that is, materials consisting of carbon atoms with the planar hexagonal networks of the graphite structure, but lacking the crystallographic order in the c direction.[1] This is a fundamental difference and these two groups of materials are distinct in many respects, with distinct properties and different applications.

As a reminder and as mentioned in Ch. 1, the term "carbon" by itself should describe the element and nothing else. To describe a material, it is coupled with a qualifier, such as "carbon black," "activated carbon," "vitreous carbon," "amorphous carbon," and others.

1.2 Structure of the Graphite Crystal

Graphite is composed of series of stacked parallel layer planes shown schematically in Fig. 3.1, with the trigonal sp^2 bonding described in Ch. 2, Sec. 4.0. In Fig. 3.1 (and subsequent figures of the carbon structure), the circles showing the position of the carbon atoms do not represent the actual size of the atom. Each atom, in fact, contacts its neighbors.

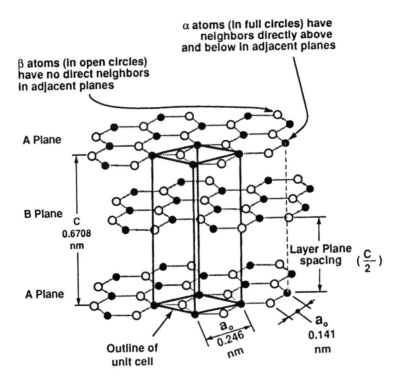

Figure 3.1. Crystal structure of graphite showing ABAB stacking sequence and unit cell.

Within each layer plane, the carbon atom is bonded to three others, forming a series of continuous hexagons in what can be considered as an essentially infinite two-dimensional molecule. The bond is covalent (sigma) and has a short length (0.141 nm) and high strength (524 kJ/mole). The hybridized fourth valence electron is paired with another delocalized electron of the adjacent plane by a much weaker *van der Waals bond* (a secondary bond arising from structural polarization) of only 7 kJ/mol (pi bond). Carbon is the only element to have this particular layered hexagonal structure.

The spacing between the layer planes is relatively large (0.335 nm) or more than twice the spacing between atoms within the basal plane and approximately twice the van der Waals radius of carbon. The stacking of these layer planes occurs in two slightly different ways: hexagonal and rhombohedral.

Hexagonal Graphite. The most common stacking sequence of the graphite crystal is hexagonal (alpha) with a -ABABAB- stacking order, in other words, where the carbon atoms in every other layer are superimposed over each other as shown in Fig. 3.1.

Atoms of the alpha type, which have neighbor atoms in the adjacent planes directly above and below, are shown with open circles. Atoms of the beta type, with no corresponding atoms in these planes, are shown with full circles. A view of the stacking sequence perpendicular to the basal plane is given in Fig. 3.2.

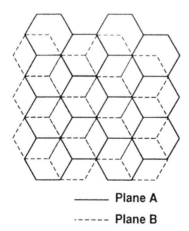

——— Plane A

------ Plane B

Figure 3.2. Schematic of hexagonal graphite crystal. View is perpendicular to basal plane.

Other characteristics of the graphite crystal are the following;

- The crystallographic description is given by the space group D^4_{6H}-$P6_3/mmc$.
- The crystal lattice parameters, i.e., the relative position of its carbon atoms (along the orthohexagonal axes) are: $a_o = 0.245$ nm and $c_o = 0.6708$ nm.
- The common crystal faces are {0002}, {1010}, {1011} and {1012}.
- The crystal cleavage is {0002} with no fracture.
- The crystal is black and gives a black streak.
- Hexagonal graphite is the thermodynamically stable form of graphite and is found in all synthetic materials.

Rhombohedral Graphite. The other graphite structure is rhombohedral with the stacking order -ABCABCABC-. The carbon atoms in every third layer are superimposed. The crystallographic description is given by the space group D^5_{3d} -$R3m$. The crystal lattice parameters are: $a_o = 0.2256$ nm and $c_o = 1.006$ nm. A view of the stacking sequence perpendicular to the basal plane is given in Fig. 3.3.

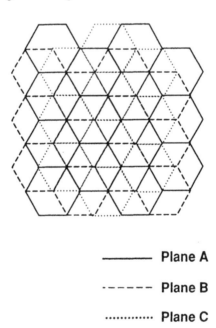

———— Plane A

– – – – – Plane B

············· Plane C

Figure 3.3. Schematic of rhombohedral graphite crystal. View is perpendicular to basal plane.

Rhombohedral graphite is thermodynamically unstable, and can be considered as an extended stacking fault of hexagonal graphite. It is never found in pure form but always in combination with hexagonal graphite, at times up to 40% in some natural and synthetic materials. It usually reverts to the hexagonal form during heat treatment above 1300°C. It should be noted that in both structures, hexagonal and rhombohedral, no basal plane lies directly over another one.

2.0   THE VARIOUS POLYCRYSTALLINE FORMS OF GRAPHITE

The ideal hexagonal graphite structure described above is composed of theoretically infinite basal planes with perfect -ABAB- stacking, with no defects whatsoever. Such an ideal structure is, of course, never found, either in natural or synthetic graphite.

2.1   Polycrystalline Graphite

Graphite materials, such as pyrolytic graphite, carbon-fiber–carbon-matrix composites (carbon-carbon), vitreous carbon, carbon black, and many others, are actually aggregates of graphite crystallites, in other words, polycrystalline graphites.[2] These crystallites may vary considerably in size. For instance, the apparent crystallite size perpendicular to the layer planes (L_c) of some vitreous carbons may be as small as 1.2 nm which is the length of a few atoms, or up to 100 nm found in highly ordered pyrolytic graphites (see Ch. 7). The layer planes may or may not be perfectly parallel to each other, depending whether the material is graphitic or non-graphitic carbon.

The aggregates of crystallites also have widely different sizes and properties. Some, such as soot, are extremely small and contain only a few small crystallites. In such cases, the properties are mostly related to the surface area (see Ch. 10).

Other aggregates may be relatively large and free of defects and essentially parallel to each other, in which case the structure and its properties closely match those of the ideal graphite crystal. Such large aggregates are often found in pyrolytic graphite (see Ch. 7).

In other aggregates, the crystallites have an essentially random orientation. This occurs in turbostratic (i.e., showing no evidence of three-dimensional order) or amorphous carbon shown in Fig. 3.4. In such cases, the bulk properties are essentially isotropic.

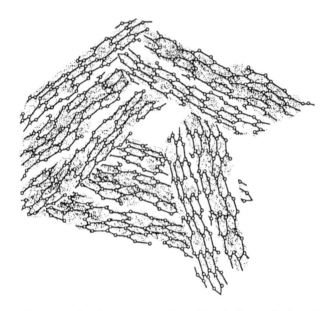

Figure 3.4. Structure of turbostratic graphite. Note lattice defects and vacancies.

2.2 Crystallite Imperfections

Within each crystallite, a varying number of imperfections may be found as shown in Figs. 3.4 and 3.5. These include:

- Vacancies, when lattice sites are unfilled indicating a missing atom within a basal plane
- Stacking faults when the ABAB sequence of the layers planes is no longer maintained
- Disclination when the planes are no longer perfectly parallel

Other crystalline imperfections likely caused by growth defects are screw dislocations and edge dislocations (Fig. 3.6). The presence of these imperfections may have a considerable influence on the properties of the bulk material.

Thus in each graphitic material, the size, shape, and degree of imperfection of the basic crystallite, the general orientation of these crystallites, as well as the bulk characteristics such as porosity and amount of impurities, may vary considerably from one material to another. As a result, the properties of these various materials may show considerable differences.

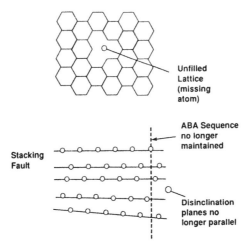

Figure 3.5. Schematic of crystallite imperfections in graphite showing unfilled lattice, stacking fault, and disinclination.

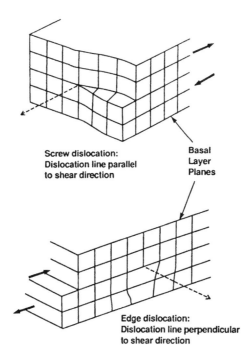

Figure 3.6. Shear dislocations in a graphite crystal.

An important implication is that, whereas material differences in the various carbon materials were originally ascribed to the presence (or absence) of an "amorphous" component (as in lampblack for instance), a more realistic approach is to relate these differences to the size and orientation of the graphite crystallites.

The specific structure and properties of the different graphitic materials will be reviewed in detail in subsequent chapters.

3.0 PHYSICAL PROPERTIES OF GRAPHITE

The properties of the ideal graphite material, that is a material that most closely corresponds to an infinitely large graphite crystal, are reviewed in this section. Such a material does not exist in the real world and the properties given below are either calculated or based on the actual properties of graphite crystals closely approaching this ideal structure.

As already mentioned and as will be seen in later chapters, a wide range of materials comes under the heading of carbon or graphite and these materials often have properties that are much different from those of the ideal graphite crystal. Obviously it is necessary to define the material accurately when speaking of the properties of "carbon" or "graphite".

3.1 Anisotropy of the Graphite Crystal

The peculiar crystal structure of graphite results in a considerable anisotropy, that is the properties of the material may vary considerably when measured along the ab directions (within the plane) or the c direction (perpendicular to the planes). Such anisotropy, especially in electrical and thermal properties, can often be put to good use as will be seen in later chapters.

3.2 Summary of Physical Properties

The physical properties of graphite are summarized in Table 3.1. It should be stressed that to obtain accurate measurements of the properties of materials much above 3000 K is a trying task. In the case of graphite, many of these measurements are based on carbon-arc experiments which are difficult to perform and interpret. The results must be viewed accordingly and some of these results and conclusions are still controversial.

Table 3.1. Physical Properties of Graphite

Crystalline form: hexagonal
Lattice parameters: a_0 = 0.246 nm
c_0 = 0.671 nm
Color: Black
Density at 300 K, 1 atm: 2.26 g/cm^3 (see below)
Atomic volume: 5.315 cm^3/mol
Sublimation point at 1 atm (estimated): 4000 K (see below)
Triple point (estimated): 4200 K (see below)
Boiling point (estimated): 4560 K
Heat of fusion: 46.84 kJ/mol
Heat of vaporization to monoatomic gas (estimated): 716.9 kJ/mol (see below)
Pauling electronegativity: 2.5

3.3 Density

The density of the perfect crystal listed in Table 3.1 is the theoretical density. Most graphite materials will have lower densities due to the presence of structural imperfections such as porosity, lattice vacancies and dislocations.

With the exception of boron nitride, graphite materials have a lower density than all other refractory materials as shown in Table 3.2. This is a advantageous characteristic especially in aerospace applications.

3.4 Melting, Sublimation, and Triple Point

The melting point of a crystalline material such as graphite is the temperature at which the solid state is in equilibrium with the liquid at a given pressure. "Normal" melting point occurs at a pressure of one atmosphere. Graphite does not have a normal melting point since, at one atmosphere, it does not melt but sublimes when the temperature reaches approximately 4000 K. To observe melting, a pressure of 100 atm and a temperature of 4200 K are necessary.

Table 3.2. Density of Some Refractory Materials

	g/cm^3
Graphite	2.26
Molybdenum	10.22
Rhenium	21.04
Tantalum	16.6
Tungsten	19.3
Titanium diboride	4.50
Hafnium carbide	12.20
Tantalum carbide	13.9
Boron nitride	2.25
Aluminum oxide	3.97
Zirconium oxide	5.89

The triple point (where all three phases, solid, liquid, and gas, are in equilibrium) is achieved, by recent estimates, at a temperature of 4200 K and a pressure of 100 atm, as shown in the vapor-pressure curve of Fig. 3.7.[3]-[5] A great deal of uncertainty still remains regarding these values of pressure and temperature, reflecting in part the difficulty of conducting experiments under such extreme conditions.

The onset of sublimation and the melting temperatures are apparently close. At temperatures above that of the triple point and at pressures of argon greater than 100 atm, a mixture of solid and liquid carbon is detected.

Graphite can thus be considered the most refractory of all the elements, tungsten being second-best with a melting point of 3680 K. However hafnium carbide (HfC) and tantalum carbide (TaC) are reported to have higher melting points (approximately 4220 K and 4270 K respectively) and are the most refractory of all materials.[6]

3.5 Heat of Vaporization

The heat of vaporization of graphite is higher than that of many metals as shown in Table 3.3.[7][8] Reported values show considerable differences.

The large amount of energy required to vaporize graphite is used to good advantage in the design of ablative structures such as nose cones and rocket nozzles (see Ch. 7 and 9).

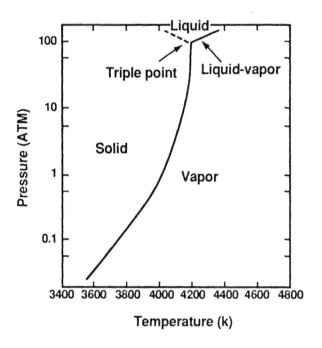

Figure 3.7. Vapor pressure and triple point of graphite.[3][4]

Table 3.3. Heat of Vaporizaton of Selected Elements at Boiling Point

	(kJ/mol)
Graphite	355.8 - 598.2
Molybdenum	598.0
Copper	300.3
Iron	349.6
Nickel	370.4
Tungsten	824.0
Silver	250.5

4.0   THERMAL PROPERTIES OF GRAPHITE

4.1   Summary of Thermal Properties

The physical properties reviewed in the previous section are essentially unaffected by the size and orientation of the crystallites in the aggregate (with the exception of density). As a result, they can be considered valid for all forms of graphite. This is no longer true for some of the properties listed in this and the following sections, and these properties may vary considerably depending on crystallite size and orientation and other factors related to the processing conditions.

The thermal properties are summarized in Table 3.4. Whenever possible, a range of property values is given. More detailed values are given in subsequent chapters.

Table 3.4. Theoretical Thermal Properties of Graphite

Heat of combustion ΔHco @ 25°C and constant pressure to form CO_2 gas, kJ/mol	393.13
Standard entropy S° at 25°C, J/mol·K	5.697 - 5.743
Entropy ΔS_{298}, J/mol·K	152.3
Enthalpy ΔH298, kJ/mol	716.88
Specific heat @ 25°C, kJ/kg·K (see below)	0.690 - 0.719
Thermal conductivity @ 25°C, W/m·K (see below)	
ab directions	398
c direction	2.2
Thermal expansion: see below	

4.2   Heat Capacity (Specific Heat)

The molar heat capacity (specific heat) of graphite is reported as 8.033 - 8.635 J/mol·K at 25°C.[8]-[10] As with all elements, it increases with temperature, with the following relationship (T in deg K).[11]

$$C_p = 4.03 + (1.14 \times 10^{-3})T - (2.04 \times 10^5)/T^2$$

The specific heat increases rapidly with temperature, up to 1500 K where it levels off at approximately 2.2 kJ/kg·K as shown in Fig. 3.8.[3] It is believed to be relatively insensitive to the differences between the various grades of synthetic graphite, and the spread of values found in the literature may be attributed to experimental variations. The average value is compared to that of selected other elements in Table 3.5. As can be seen, it is higher than most metals.

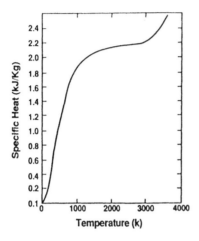

Figure 3.8. Specific heat of graphite vs. temperature at one atmosphere.[3]

Table 3.5. Specific Heat of Selected Elements

	C_p at 25°C and 1 atm. (kJ/kg·K)
Graphite	0.690 - 0.719
Diamond	0.502 - 0.519
Boron	1.025
Aluminum	0.900
Titanium	0.523
Copper	0.385
Niobium	0.263
Rhenium	0.242
Tungsten	0.130
Water	4.186

4.3 Thermal Conductivity

The thermal properties of conductivity and expansion are strongly influenced by the anisotropy of the graphite crystal. The thermal conductivity (K) is the time rate of transfer of heat by conduction. In graphite, it occurs essentially by lattice vibration and is represented by the following relationship (Debye equation):

Eq. (1) $K = bC_pvL$

where: b = a constant
C = specific heat per unit volume of the crystal
v = speed of heat-transporting acoustic wave (phonon)
L = mean free path for wave scattering

In a polycrystalline materials, the waves (phonon, i.e., quantum of thermal energy) are scattered by crystallite boundaries, lattice defects, and other phonons. Little of this occurs in a perfect or near-perfect graphite crystal in the basal plane and, as a result, the factor L is high and thermal conductivity is high in the *ab* directions. However, in the direction perpendicular to the basal plane (*c* direction), the conductivity is approximately 200 times lower since the amplitude of the lattice vibration in that direction is considerably lower than in the *ab* directions. These differences in vibration amplitude in the various crystallographic directions of graphite are shown in Fig. 3.9.[12]

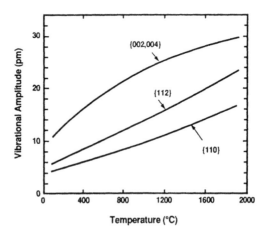

Figure 3.9. Thermal vibrational amplitude of graphite crystal in the {002,004}, {112}, and {110} directions.[12]

The thermal conductivity of a graphite crystal has been reported as high as 4180 W/m·K in the *ab* directions for highly crystalline, stress-annealed pyrolytic graphite.[2] However, the average value for commercial pyrolytic graphite is considerably smaller (~390 W/m·K).[13] Still, this is a high value and graphite, in the *ab* directions, can be considered a good thermal conductor comparable to high-conductivity metals and ceramics as shown in Table 3.6. Graphite fibers from pitch precursor have high thermal conductivity up to 1180 W/m·K, nearly three times that of copper (see Ch. 8, Sec. 6).

Table 3.6. Thermal Conductivity of Selected Materials

	W/m·K at 25°C
Pyrolytic graphite:	
ab directions	390
c direction	2
Graphite fiber (pitch-based)	1180
Diamond (Type II)	2000 - 2100
Silver	420
Copper	385
Beryllium oxide	260
Aluminum nitride	200
Alumina	25

The thermal conductivity in the *c* direction is approximately 2.0 W/m·K and, in that direction, graphite is a good thermal insulator, comparable to phenolic plastic.

The thermal conductivity of graphite decreases with temperature as shown in Fig. 3.10.[14] In the Debye equation (Eq. 1), K is directly proportional to the mean free path, L, which in turn is inversely proportional to temperature due to the increase in vibration amplitude of the thermally excited carbon atoms. L becomes the dominant factor above room temperature, more than offsetting the increase in specific heat, C_p, shown in Fig. 3.8.

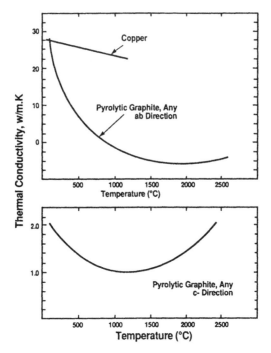

Figure 3.10. Thermal conductivity of graphite crystal in the *ab* and *c* directions.[14]

4.4   Thermal Expansion

The interatomic spacing between the carbon atoms of graphite (as with any other material) is a function of temperature. At 0 K (-273°C), these atoms have their lowest energy position or *ground state* (see Ch. 2, Sec. 4). The increased energy resulting from increasing temperature causes the atoms to vibrate and move further apart. In other words, the mean interatomic spacing increases and the result is thermal expansion.

This can be represented graphically in Fig. 3.11. As seen in this figure, the graphic relationship between interatomic spacing and energy has the configuration of a trough. This configuration changes with the strength of the atomic bond. In a strongly bonded solid such as graphite in the *ab* directions, the trough is deep, the amplitude of the vibrations is small and, during the outward motion of the atoms, the atomic bonds are not overstretched and, consequently, the dimensional changes remain small. When the atomic bond is weak such as in graphite in the *c* direction, the energy trough is shallow and the vibration amplitude and the dimensional changes are large.[15]

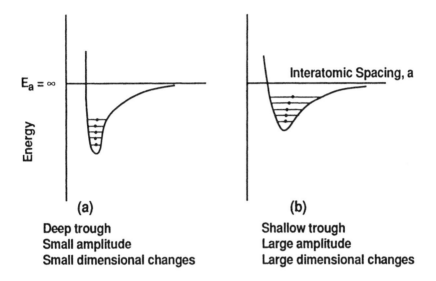

Figure 3.11. The energy trough of graphite in *(a) ab* directions and *(b) c* direction.[14]

As a result, the thermal expansion of the graphite crystal has a marked anisotropy. It is low in the *ab* directions (lower than most materials) but higher by an order of magnitude in the *c* direction, as shown in Fig. 3.12.[2][13][16]

The increase with temperature is not linear. In the *c* direction, it increases slowly and gradually. At 0°C, the *coefficient of thermal expansion* averages 25 x 10⁻⁶/°C and at 400°C, it reaches 28 x 10⁻⁶/°C.[2][14][17]

In the *ab* directions, the thermal expansion is actually negative up to approximately 400°C with a minimum at 0°C. It is possible that this observed negative expansion is due to internal stress (Poisson effect) associated with the large expansion in the *c* direction and it has been suggested that, if it were possible to measure the *ab* thermal expansion of a single atomic plane, this expansion would be positive.[18]

The large thermal expansion anisotropy often results in large internal stresses and structural problems such as delamination between planes as will be seen in Ch. 5, Sec. 3.

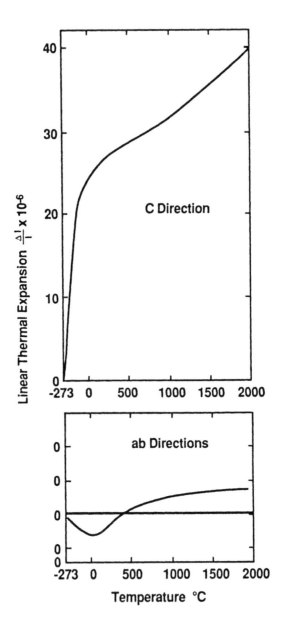

Figure 3.12. Thermal expansion of the graphite crystal in the *ab* and *c* directions.[2][3][16]

5.0 ELECTRICAL PROPERTIES OF GRAPHITE

5.1 Electrical Resistivity

In electrical conductors such as metals, the attraction between the outer electrons and the nucleus of the atom is weak; the outer electrons can move readily and, since an electric current is essentially a flow of electrons, metals are good conductors of electricity. In electrical insulators (or dielectrics), electrons are strongly bonded to the nucleus and are not free to move.[18][19]

Electrically, graphite can be considered as a semi-metal, that is a conductor in the basal plane and an insulator normal to the basal plane. Its atomic structure is such that the highest-filled valence band overlaps the lowest-empty conduction band by approximately 36 meV and the delocalized fourth-valence electrons form a partially-filled conduction band between the basal planes where they can moved readily in a wave pattern as they respond to electric fields.[20] Consequently, the electrical resistivity of graphite parallel to the basal planes (*ab* directions) is low and the material is a relatively good conductor of electricity.

In the *c* direction, the spacing between planes is comparatively large, and there is no comparable mechanism for the electrons to move from one plane to another, in other words, normal to the basal plane. As a result the electrical resistivity in that direction is high and the material is considered an electrical insulator. In some cases, it may be 10,000 times higher than in the *ab* directions.[20] Often quoted resistivity values are 3000 x 10^{-6} ohm.m in the *c* direction and 2.5 - 5.0 x 10^{-6} ohm.m in the *ab* directions.

5.2 Resistivity and Temperature

The electrical resistivity of the graphite crystal in the *ab* directions increases with temperature, as does that of metals. This increase is the result of the decrease in the electron mean free path, in a mechanism similar to the increase in thermal conductivity reviewed above in Sec. 4.3.

The electrical resistivity in the *c* direction, however, decreases slightly with increasing temperature, possibly because electrons can jump or tunnel from one plane to another due to increased thermal activation.[14]

6.0   MECHANICAL PROPERTIES OF GRAPHITE

As mentioned in Sec. 1.2, the bond between atoms within the basal plane of a graphite crystal is considerably stronger than the bond between the planes with an anisotropy ratio of approximately 75. This means that, while the strength in the *ab* directions is considerable, that in the *c* direction (interlaminar strength) is low and graphite shears easily between basal planes. The elastic constants are:[17]

C_{11} = 1060 GPa (*a* direction)

C_{33} = 36.5 GPa (*c* direction)

C_{44} = 4.5 GPa (parallel to the planes)

The Young's modulus of elasticity of the crystal varies up to two orders of magnitude with the direction. It is plotted in Fig. 3.13 as a function of the angle between the *c* direction and the direction of measurement.[2]

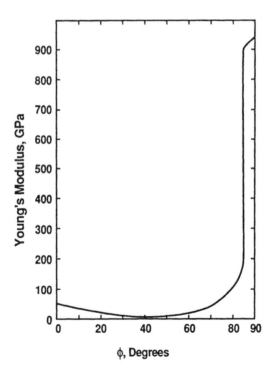

Figure 3.13. Young's modulus of graphite crystal as a function of angle (Θ) with *c* axis.[2]

It should be noted that such values are those of the ideal graphite crystal. The mechanical properties of "real" material, such as the molded graphite materials and pyrolytic graphite are considerably different since they are controlled by the dominant crystallite orientation, porosity, structural defects and other factors. The result is large variations in the range of these properties, depending on type of graphite and process. The mechanical properties of the various forms of graphite will be reviewed in detail in subsequent chapters.

7.0 CHEMICAL PROPERTIES

7.1 General Considerations

Pure graphite is one of the most chemically inert materials. It is resistant to most acids, alkalies and corrosive gases. However impurities are almost always present to some degree in both natural and artificial graphites and often have an important catalytic effect with resulting increase in the chemical reactivity.

The anisotropy of the graphite crystal is reflected in its chemical behavior. Reaction with gases or vapors occurs preferentially at "active sites", i.e., the end of the basal planes of the crystal which are the zigzag face {101} and the arm-chair face {112} as shown in Fig. 3.14, and at defect sites, such as dislocations, vacancies, and steps. Reaction with the basal plane surfaces is far slower. The reason is that the graphite crystal exhibits large differences in surface energy in the different crystallographic directions; these energies amount to 5 J/m^2 in the prismatic plane but only 0.11 J/m^2 in the basal plane. These differences account for the different rate of reaction, i.e., slow at the basal plane and rapid at the edge (or prismatic) surfaces found at the termination of the basal planes or at defects within the basal plane.[21] Consequently, graphite materials with large crystals and few defects have the best chemical resistance.

The chemical reactivity is also appreciably affected by the degree of porosity, since high porosity leads to large increase in surface area with resulting increase in reactivity. Differences in reactivity between one form of graphite or another can be considerable. Obviously, high surface area materials such as activated carbon are far more reactive than dense, pore-free or closed-pore materials such as glassy carbon.

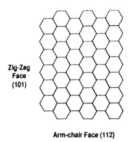

Arm-chair Face (112)

Figure 3.14. The faces of a graphite crystal.

Reactivity also generally increases with increasing temperature and, at high temperatures, graphite becomes far more reactive. For instance, above 450°C, it oxidizes readily with water, oxygen, some oxides, and other substances.

In this section, only the general chemical behavior of graphite will be considered. Reviews of the chemical reactivity of specific graphite materials will be given in subsequent chapters.

7.2 Summary of Chemical Properties

The general reactivity of graphite with the various classes of chemicals at ambient temperature is given in Table 3.7.[22]

7.3 Reaction with Oxygen and Hydrogen

The one major exception to the generally excellent chemical resistance of graphite is poor resistance to the elements of Column VI (oxygen, sulfur, selenium and tellurium), particularly oxygen and oxygen compounds. Oxidation begins in air at 350 - 400°C. This low-temperature oxidation is in contrast with the behavior of other refractory materials: oxides, for instance, do not oxidize and many carbides form a protective oxide film on their surface that delays oxidation. In contrast, the oxides formed by the oxidation of graphite are gaseous (CO and CO_2) and offer no protection to the surface.

As mentioned in Sec. 7.1 above, the reaction rate is site preferential and oxidation is much higher along the zig-zag face of the crystal than it is along the armchair face (Fig. 3.14).[23][24] Oxidation can be hindered by increasing the degree of graphitization and the crystallite size (thereby reducing the number of active sites).

Table 3.7. General Chemical Resistance of Graphite

Chemical	Resistance to Chemical Attack
Acids and Acid Solutions	
• Mineral, non-oxidizing (HCl, HCN, H_3PO_4, HF)	A
• Boiling H_2SO_4	B
• Mineral, oxidizing (Br_2, $H_2Cr_2O_7$, HNO_2, $HClO_4$)	B
• Inorganic salts, acid-forming (alum, BF_3, $CuCl_2$, $NiCl_2$, sulfates)	A
• Organic, strong (pH <3) (acetic, carbolic, formic, maleic, oxalic, phenol, picric, salicic)	A
• Organic, weak (pH 3 - 7) (abietic, benzoic, citric, cresol, lactic, palmitic)	A
• Organic salts, acid-forming (allyl, amyl chlorides, ethyl chlorides)	A
Alkalis and Alkaline Solutions	
• Mineral, non-oxidizing (hydroxides, NaOH, hydrazine, molten Mg, KOH)	A
• Mineral, oxidizing (permanganate, perchlorate, perborates)	B
• Inorganic salts, base forming ($BaSH_2$, borax, phosphates, NH_2O_8, $NaHSO_2$, $NaPO_3$, Na_2S)	A
• Organic, strong (pH >11) (ethanol amines, pyridine)	A
• Weak organic bases (pH 7 - 11) (aniline, soaps, urea)	A
Gases	
• Acid (BF_3, CO_2, Cl_2, HCl, HF, H_2S)	A
• Alkaline (wet NH_3, steam to 300°C)	A
• Anhydrous (dew point below 0°C) (NH_3, CO_2, ethane, F_2, H_2, HCl, HF, H_2S, methane, O_2 to 150°C, propane, SO_2)	B
• Liquefied (air, F_2, He, H_2, methane, N_2, O_2)	C
• Oxidizing (air above 250°C, F_2, N_2O_4, O_2 above 150°C, steam above 300°C)	C
• Reducing (acetylene, ethane, methane)	A

Melts

- Acid salts ($AlCl_3$, H_2BO_3, $FeCl_3$, PCL_3, $ZnCl_2$)   B
- Alkaline salts ($Ba(OH)_2$, LiOH, KCN, soda ash)   B
- Metals (Al, Sb, Babbit, brass, Cu, Ga, Au, Mg, Hg, Ag, Sn, Zn)   A
- Neutral salts (KCl, Na_2SO_4)   A
- Oxidizing salts (sodium nitrate)   B
- Salt solutions, neutral (baking soda, $KCr(SO_4)_2$, $CuSO_4$, $Mg(SO_4)_2$, KCL, sea water, sewage, Na_2SO_4)   A

Solvents

- Aliphatic (butadiene, butane, butylene, cyclohexane, fuel oil, gasoline, lubric oil, propane, propylene)   A
- Aromatic (benzene, coal tar, creosote, cumene, naphtalene, petroleum, styrene, urethane)   A
- Chlorinated, fluorinated (CCl_4, chlorobenzene, freons, chloroform, methyl chloride, vinyl chloride)   A
- Oxygenated, sulfide (acetaldehyde, acrolein, butyl acetate, butyl alcohol, CS_2, rayon, ether, ethyl acetate, furfural, glycerine, methanol, ketones, sorbitol, vinyl acetate)   A

(A=high, B=medium, C=low)

The oxidation of graphite and the available protective coatings are reviewed in Ch. 9. The controlled oxidation of graphite, known as activation, results in open structures with extremely high surface area (see Ch. 5, Sec. 3.0).

Graphite does not react with hydrogen at ordinary temperatures. It reacts in the 1000 - 1500°C range to form methane (CH_4). The reaction is accelerated in the presence of a platinum catalyst. With nickel catalyst, the reaction begins at approximately 500°C.[25]

7.4 Reaction with Metals

Graphite reacts with metals that form carbides readily such as the metal of groups IV, V and VI.[12][25] These carbides are the so-called hard carbides, which include the carbides of tungsten, molybdenum, titanium, vanadium and tantalum, as well as the non-metal carbides of silicon and boron. Graphite reacts with iron to form iron carbide, Fe_3C, usually by the direct solution of carbon in the molten iron. Iron carbide may also be formed from the reaction of iron with a carbon-containing gas. This reaction is known as case-hardening.

The reaction rate of graphite with the precious metals, aluminum, and the III-V and II-VI semiconductor compounds is low and graphite is used successfully as a crucible to melt these materials.

Graphite reacts readily with the alkali metals: potassium, calcium, strontium, and barium. The atoms of some of these metals, notably potassium, can readily penetrate between the basal planes of the graphite crystal to form intercalated (or lamellar compounds) with useful properties. These compounds are reviewed in Ch. 10, Sec. 3.0.

7.5   Reaction with Halogens, Acids, and Alkalis

Like the alkali metals, some halogens, particularly fluorine, form intercalated compounds with graphite crystals. Reaction usually starts at 600°C. However, graphite does not react with chlorine at temperatures below that of the electric arc.

Oxidizing acids attack graphite to varying degree depending on the nature and surface area of the material. The reaction with concentrated nitric acid is as follows:

$$C + 4HNO_3 \rightarrow 2H_2O + 4NO_2 + CO_2$$

Depending on the reaction conditions, other products may be formed such as graphitic oxide ($C_7H_2O_4$), mellitic acid ($C_6(CO_2H)_6$) and hydrocyanic acid (HCN).[7]

Another oxidizing acid that attacks graphite is boiling sulfuric acid. The simplified reaction is the following:

$$C + 2H_2SO_4 \rightarrow CO_2 + 2H_2O + 2SO_2$$

Other by-products may be formed such as benzoic acid, $C_6H_5CO_2H$, and mellitic acid, $C_6(CO_2H)_6$.

Hydrofluoric acid (HF) and the alkali hydroxides generally do not react with graphite.

REFERENCES

1. "International Committee for Characterization and Terminology of Carbon," *Carbon*, 28(5):445-449 (1990)

2. Bokros, J. C., in *Chemistry and Physics of Carbon*, (P. L. Walker, Jr., ed.), Vol. 5, Marcel Dekker Inc., New York (1969)

3. Kohl, W. H., *Handbook of Materials and Techniques for Vacuum Devices*, Reinhold Publishing, New York (1967)

4. Palmer, H. B. and Shelef, M., in *Chemistry and Physics of Carbon*, (P. L. Walker, Jr., ed.), Vol. 4, Marcel Dekker Inc., New York (1968)

5. Gustafson, P., *Carbon*, 24(2)169-176 (1986)

6. Storms, E. K., *The Refractory Carbides*, Academic Press, New York (1968)

7. Mantell, C. L., *Carbon and Graphite Handbook*, Interscience Publishers, New York, (1968)

8. Perry's *Chemical Engineering Handbook*, 6th ed., McGraw-Hill, New York (1984)

9. Wehr, M. R., Richards, J. A. Jr., and Adair, T. W. III, *Physics of the Atom, 3rd ed.*, Addison-Wesley Publishing, Reading, MA (1978)

10. *Chart of the Atoms*, Sargent-Welch Scientific Co., Skokie, IL (1982)

11. Eggers, D. F., Gregory, N. W., Halsey, G. D., Jr., and Rabinovitch, B. S., *Physical Chemistry*, John Wiley & Sons, New York (1964)

12. Fitzer, E., *Carbon*, 25(2):1633-190 (1987)

13. Graphite, Refractory Material, Bulletin from *Le Carbone-Lorraine*, Gennevilliers 92231, France

14. Campbell, I. E. and Sherwood, E. M., *High-Temperature Materials and Technology*, John Wiley & Sons, New York (1967)

15. Van Vlack, L. H., *Elements of Materials Science and Engineering*, Addison-Wesley Publishing Co., Reading, MA (1980)

16. Sze, S. M., *Semiconductor Physics and Technology*, John Wiley & Sons, New York (1985)

17. Fitzer, E., *Carbon*, 27(5):621-645 (1989)

18. Mullendore, A. M., Sandia Park NM, Private Communication (1992); Nelson, and Riley, *Proc. Phys. Soc.*, London 57:477 (1945)

19. Murray, R. L. and Cobb, G. C., *Physics, Concepts and Consequences*, Prentice Hall Inc., Englewood Cliffs, NJ (1970)

20. Spain, I. L., in *Chemistry and Physics of Carbon,* (P. L. Walker and P. A. Thrower, eds.), Vol. 8, Marcel Dekker Inc., New York (1973)

21. Walker, P. L., Jr., *Carbon,* 28(3-4):261-279 (1990)

22. *Corrosion/Chemical Compatibility Tables,* Bulletin of the Pure Carbon Co., St. Marys, PA (1984)

23. Hippo, E. J., Murdie, N., and Hyjaze, A., *Carbon,* 27(5):689-695 (1989)

24. Yavorsky, I. A. and Malanov, M. D., *Carbon,* 7:287-291 (1989)

25. *Carbon/Graphite Properties,* Bulletin from The Stackpole Carbon Co., St. Marys, PA (1987)

4

Synthetic Carbon and Graphite: Carbonization and Graphitization

1.0 TYPES OF SYNTHETIC CARBON AND GRAPHITE

Chapters 2 and 3 were a review of the carbon atom and its bonding mechanisms and how these atoms combine to form graphite crystals. In this and the next six chapters, the focus will be on how large numbers of these crystallites are combined to form synthetic (and natural) carbon and graphite products. The various types of synthetic materials will be reviewed including their production processes, their properties and characteristics, and their present and potential applications.

In terms of size, the review proceeds from the size of a single carbon atom, to that of a graphite crystal, composed of thousands of atoms, to that of a graphite product such as an electrode, which may weigh hundreds of kilograms.

Natural graphite, which is found in abundance in many areas of the world, has been used since historical times, but its applications always were (and still are) limited (see Ch. 10). In the last century, the advent of synthetic graphite and carbon has considerably increased the scope of applications, although natural graphite still remains the material of choice in a few cases. A large majority of graphite and carbon products are now synthetic and these products are continuously being improved and upgraded.

1.1 Synthetic Graphite and Carbon Products

The carbon terminology was reviewed in Ch. 3, Sec. 1.1, and its proper use is important as confusion can easily prevail because of the many variations of graphite and carbon products. The synthetic graphite and carbon products can be divided into six major categories:

1. Molded graphite and carbon (Ch. 5)
2. Vitreous (glassy) carbon (Ch. 6)
3. Pyrolytic graphite and carbon (Ch. 7)
4. Carbon fibers (Ch. 8)
5. Carbon composites and carbon-carbon (Ch. 9)
6. Carbon and graphite powders and particles (Ch. 10)

These divisions may appear arbitrary and overlapping in some cases; for instance, fibers and carbon-carbon are generally polymeric carbons, although pyrolytic graphite is often used in their processing. These divisions however correspond to specific and unique processes, with resulting products that may have different properties. These variations in properties, as stated in Ch. 3 (Sec. 2.1), stem from the nature of the polycrystalline aggregates, their different crystallite sizes and orientation, various degrees of porosity and purity, and other characteristics.

1.2 General Characteristics of Synthetic Graphite and Carbon

Many new graphite- and carbon-materials with improved characteristics have been developed in the last two decades. Some of these materials have a strongly anisotropic structure and properties that approach those of the perfect graphite crystal. Others have a lesser degree of anisotropy which is not always a disadvantage as, in many cases, isotropic properties are a desirable feature, as will be seen in later chapters.

A common characteristic of graphite and carbon materials, whatever their origin or processing, is that they are all derived from organic precursors: molded graphite from petroleum coke and coal-tar, pyrolytic graphite from methane and other gaseous hydrocarbons, vitreous carbon and fibers from polymers, carbon black from natural gas, charcoal from wood, coal from plants, etc.

These organic precursors must be carbonized and, more often than not, graphitized, in order to form carbon and graphite materials. The critical and complex phenomena of carbonization and graphitization are the two common features of the production of all these synthetic materials with the notable exception of pyrolytic graphite, which is produced by the entirely different process of vapor deposition (reviewed in Ch. 7). These two phenomena, carbonization and graphitization, are the topics of this chapter.

2.0 THE CARBONIZATION (PYROLYSIS) PROCESS

The carbonization process, also known as *pyrolysis*, can be defined as the step in which the organic precursor is transformed into a material that is essentially all carbon. The mechanism of carbonization is reviewed below in general terms. Additional information on the carbonization of specific materials is given in subsequent chapters.

2.1 Principle of Carbonization

Carbonization Cycle. Carbonization is basically a heating cycle. The precursor is heated slowly in a reducing or inert environment, over a range of temperature that varies with the nature of the particular precursor and may extend to 1300°C. The organic material is decomposed into a carbon residue and volatile compounds diffuse out to the atmosphere. The process is complex and several reactions may take place at the same time such as dehydrogenation, *condensation* and *isomerization*.

The carbon content of the residue is a function of the nature of the precursor and the pyrolysis temperature. It usually exceeds 90 weight % at 900°C and 99 weight % at 1300°C.

The diffusion of the volatile compounds to the atmosphere is a critical step and must occur slowly to avoid disruption and rupture of the carbon network. As a result, carbonization is usually a slow process. Its duration may vary considerably, depending on the composition of the end-product, the type of precursor, the thickness of the material, and other factors. Some carbonization cycles, such as those used in the production of large electrodes or some carbon-carbon parts, last several weeks. Others are considerably shorter, such as the carbonization cycle to produce carbon fibers, since these fibers have a small cross-section and the diffusion path is short. The specifics of each cycle will be reviewed in more detail in the following chapters.

Characteristics of Carbonized Materials. After carbonization, the residual material is essentially all carbon. However, its structure has little graphitic order and consists of an aggregate of small crystallites, each formed of a few graphite layer planes with some degree of parallelism and usually with many imperfections. These crystallites are generally randomly oriented as described in Ch. 3, Sec. 2.0 and shown in Fig. 3.4.

The carbonized material is often called "amorphous" or "baked carbon". It is without long-range crystalline order and the deviation of the interatomic distances of the carbon atoms (from the perfect graphite crystal) is greater than 5% in both the basal plane (*ab* directions) and between planes (*c* direction), as determined by x-ray diffraction.

Amorphous carbon is hard, abrasion resistant, brittle, and has low thermal- and electrical-conductivities. In a few cases, these characteristics are desirable and amorphous carbon is found in applications such as contacts, pantographs, current collectors and brushes for operation on flush mica commutators, as well as in special types of carbon-carbon.[1]

In most instances however, amorphous carbon is only the intermediate stage in the manufacture of synthetic graphite products.

2.2   Precursor Materials and Their Carbon Yield

The carbon yield is defined as the ratio of the weight of the carbon residue after carbonization to the weight of the material prior to carbonization.

Typical carbon yields of common and potential precursor materials are shown in Table 4.1.[2][3] These yields are not fixed but depend to a great degree on the heating rate, the composition of the atmosphere, the pressure, and other factors (see below). The nature of the carbon yield, given in the last column, i.e., coke or char, is reviewed in the following section on graphitization.

Effect of Pressure on Carbon Yield and Structure: The nature and the length of the carbonization cycle are important factors in controlling the carbon yield. For instance, the effect of gas pressure can be considerable. Figure 4.1 shows this effect on the yield of three grades of coal-tar pitch, with various softening points.[4] In this particular case, high pressure more than doubles the yield. Pressure can also modify the structure of the resulting carbon and change its graphitization characteristics.[5]

Table 4.1. Typical Carbon Yield of Various Precursor Materials

Precursor	Average carbon yield (%)	Type of carbon*
Aromatic hydrocarbons		
Coal-tar pitches	40 - 60	Coke
Petroleum fractions	50 - 60	Coke
Naphtalene, $C_{10}H_8$		Coke
Anthracene, $C_{14}H_{10}$		Coke
Acenaphtalene, $C_{12}H_8$		Coke
Phenantrene, $C_{14}H_{10}$		Char
Biphenyl, $C_{12}H_{10}$		Char
Polymers		
Polyvinyl chloride, $(CH_2CHCl)_n$	42	Coke
Polyimide (Kapton), $(C_{22}H_{10}O_5N_2)_n$	60	Coke
Polyvinylidene chloride, $(CH_2CCl_2)_n$	25	Char
Polyfurfural alcohol, $(C_5O_2H_6)_n$	50 - 56	Char
Phenolics, $(C_{15}O_2H_{20})_n$	52 - 62	Char
Polyacrylonitrile (PAN), $(CH_2CHCN)_n$	46 - 50	Char
Cellulose, $(C_{12}O_{10}H_{18})_n$	20	Char

* Coke is a graphitizable carbon, char is non-graphitizable (see Sec. 3.0 below).

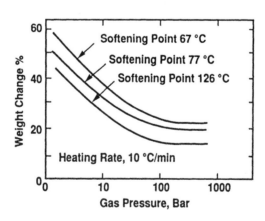

Figure 4.1. Effect of gas pressure on weight change during pyrolysis of various pitches at 600°C.[4]

2.3   Carbonization Mechanism of Aromatic Hydrocarbons

As shown in Table 4.1, the graphite precursors can be divided into two major classes: *(a) aromatic hydrocarbons* and *(b) polymers*, each with different carbonization characteristics.

Structure of Aromatic Hydrocarbons. The term hydrocarbon refers to an organic compound that contains only carbon and hydrogen. Aromatics are hydrocarbons characterized by the presence of at least one *benzene ring*. Aromatics have a graphite-like structure and graphite is often considered as the parent of all these compounds. The structure of benzene is shown below:

The structures of the aromatics listed in Table 4.1 are shown in Figs. 4.2 (coke formers) and 4.3 (char formers). Some of the most important aromatics are the following:[6]

- Anthracene is a linear, planar molecule with three benzene rings. In an autoclave at approximately 450°C, it begins to lose the hydrogen atoms in the 9,10 positions. Free radicals are formed and condensed into gradually larger, planar molecules and eventually coke is formed.

- Phenanthrene is a branched, planar *isomer* of anthracene. It carbonizes to a char in a manner similar to anthracene but with a lower yield.

- Biphenyl has two benzene rings connected by a single carbon-carbon bond. It is non-planar with free rotation around this bond. It carbonizes to a char.

Mesophase. The general carbonization mechanism of polyaromatic hydrocarbons is relatively simple, at least in theory, since it proceeds by the rupturing of the carbon-hydrogen bonds and the removal of the hydrogen.[2] Some of these hydrocarbons first go through an intermediate liquid or plastic stage which occurs at temperatures above approximately 400°C. This stage is the so-called "mesophase", in which the material shows the optical birefringence characteristic of disk-like, nematic liquid crystals, that is, crystals which have a lamellar arrangement with the long axes in parallel lines (the so-called Brooks and Taylor morphology).[7]-[9]

During this melt stage, condensation takes place and large polyaromatic molecules are formed with a molecular weight averaging 1000. These polycyclic, spherulitic liquid crystals gradually increase in size to build up sufficient mutual *van der Waals* attraction to start promoting their alignment and form pre-cokes or "green cokes".[3] The green cokes still retain from 6 to 20% volatile matter at 600°C. These volatiles are gradually removed as the temperature is raised.

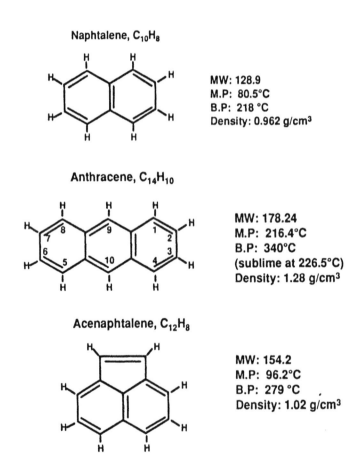

Naphtalene, $C_{10}H_8$

MW: 128.9
M.P: 80.5°C
B.P: 218 °C
Density: 0.962 g/cm³

Anthracene, $C_{14}H_{10}$

MW: 178.24
M.P: 216.4°C
B.P: 340°C
(sublime at 226.5°C)
Density: 1.28 g/cm³

Acenaphtalene, $C_{12}H_8$

MW: 154.2
M.P: 96.2°C
B.P: 279 °C
Density: 1.02 g/cm³

Figure 4.2. Structure and properties of coke-forming aromatic hydrocarbons.

Phenanthrene, $C_{14}H_{10}$

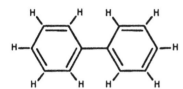

MW: 178.24
M.P: 101°C
B.P: 340 °C
(sublime at 210°C)
Density: 0.98 g/cm³

Biphenyl, $C_{12}H_{10}$

MW: 154.21
M.P: 71°C
B.P: 255°C
(sublime at 145°C)
Density: 0.86 g/cm³

Figure 4.3. Structure and properties of char-forming aromatic hydrocarbons.

Mesophase pitches with a highly oriented structure, high purity, low viscosity, and high coke yield are the precursor materials in the production of pitch-based carbon fibers and are reviewed in Ch. 8, Sec. 3.[10]

Carbonization Mechanism. The carbonization mechanism of polyaromatic hydrocarbons (PAH) includes the following steps:[11]

1. Two PAH molecules disproportionate into one hydroaromatic molecule and one free-radical molecule.

2. The free-radical molecules condense into larger molecular weight aromatics.

3. The hydroaromatic molecules cleave to produce liquid and gas-phase species.

Coal-tar pitch and petroleum pitch are complex mixtures of PAH and are widely used in the molded graphite industry. Coal-tar pitch is the solid residue remaining after the distillation of coal-tar. It is predominantly composed of aromatic- and heterocyclic-hydrocarbon compounds. Its composition varies depending on the nature of the coal-tar. It is a solid at room temperature and has a broad softening range (up to 150°C) and no well-defined melting point. The carbon yield is not fixed and varies with the composition (see Ch. 5, Sec. 1.2).

Petroleum pitch is the residue from the distillation of petroleum fractions. Like coal-tar pitch, it has a varying composition, consisting mostly of aromatic and alkyl-substituted aromatic hydrocarbons. A solid at room temperature, it also has a broad softening range and no well-defined melting point. The formation of its mesophase is related to the chemical constituents and the asphaltene fractions.[12] Coal-tar and petroleum-pitches do not crystallize during cooling but can be considered as supercooled liquids.

Single hydrocarbons of interest as precursor materials are anthracene, phenantrene and naphthalene. The latter is a precursor in the production of carbon fibers. These compounds are obtained by recovery from the distillation of coal-tar between 170 - 230°C for naphtalene, and >270°C for anthracene and phenantrene. Other coal-tar derivatives are biphenyl and truxene (see Sec. 3.2 below).

These materials can be processed singly or modified by the addition of solvents or other materials to alter their carbonization and graphitization characteristics, as will be shown in Ch. 8.[13]

2.4 Carbonization of Polymers

Polymers are organic materials consisting of macromolecules, composed of many repeating units called "mers", from which the word polymer is derived. The major polymers in the production of synthetic graphite are listed in Table 4.1 above and their chemical structures are shown in Fig. 4.4.

As a rule, polymers have a lower carbon yield than aromatic hydrocarbons since they contain, in addition to hydrogen and carbon, other elements with higher molecular weight, such as chlorine, oxygen, or nitrogen. These elements must be removed. The carbonization mechanism of polymers is usually more intricate than that of aromatic hydrocarbons since the composition is more complex. The carbon yield is unpredictable in many cases.[2]

One of the oldest polymer precursors is cellulose which has been used for generations in the production of charcoal from wood and, early in the twentieth century, for lamp filaments from cotton or bamboo by Thomas Edison.

In linear polymers (that is non-aromatic), such as polyethylene and polystyrene, the polymeric chain breaks down into increasingly smaller molecules which eventually gasify. As a result, these materials have a low carbon yield and cannot be considered as suitable precursors. In fact, gasification is sometimes almost complete and practically no carbon residue is left.

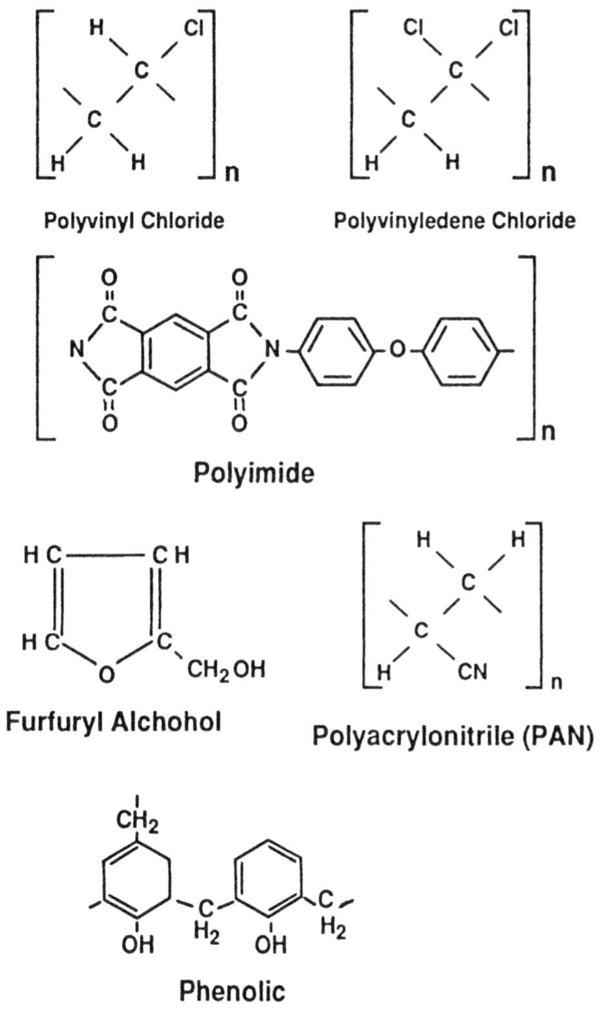

Figure 4.4. Structure of polymeric precursors commonly used in the production of synthetic graphite.

Other polymers, such as polyacrylonitrile (PAN), furfuryl alcohol, and phenolics, which have high molecular weight and a high degree of aromaticity (i.e., with many benzene rings), have a relatively high carbon yield. During carbonization, the polymeric chains do not break down and, unlike the aromatic hydrocarbons, do not go through a liquid or plastic stage. An exception to this is polyvinyl chloride (PVC) which carbonizes like an aromatic hydrocarbon. The carbonization sequence of specific polymers is described in subsequent chapters.

3.0   THE GRAPHITIZATION PROCESS

3.1   X-Ray Diffraction of Graphitic Materials

X-ray diffraction is a useful analytical technique to determine the changes in structure that occur during graphitization. A detailed analysis of this technique is given in Ref. 14. As with all crystalline materials, a sharp diffraction pattern is obtained with single-crystal graphite. This pattern is schematically shown in Fig. 4.5.[2] Pronounced crystallinity is indicated by the development of the 002, 004 and 101 peaks.

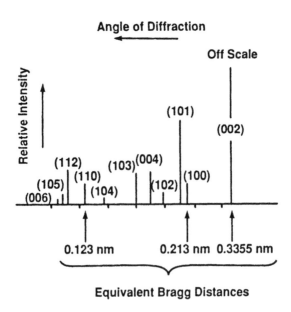

Figure 4.5. Schematic of diffraction patterns of hexagonal graphite.[2]

The crystallite size, reviewed in Ch. 3, Sec. 2.1, is measured by x-ray diffraction from the breadth of the (110) and the (002) lines. The calculated value is not the actual size of the crystallite but is a relative value which is useful in determining the degree of graphitization of a carbon structure. The interlayer spacing is also determined from x-ray measurements and is another indication of the degree of graphitization.

3.2   Coke and Char

Graphitization can be defined as the transformation of a turbostratic-graphitic material (i.e., a "carbon") into a well-ordered graphitic structure. This occurs during heat treatment at temperatures often in excess of 2500°C.[2][4] This structural change is accompanied by a large increase in the electrical and thermal conductivities of the material, an increase in its density which now approaches that of single crystal graphite, and a decrease in its hardness which facilitates machining into a finished product.

The degree of graphitization of carbon precursors such as those shown in Table 4.1 varies considerably, depending on whether a coke or a char is formed, as will be shown in the following sections.

3.3   Graphitization of Coke-Former Hydrocarbons

Cokes are formed by the carbonization of most aromatic hydrocarbons and a few polymers as mentioned above and are readily converted into a well-ordered graphite structure.

The coke-former aromatic hydrocarbons listed in Table 4.1 and shown in Fig. 4.2, i.e., naphthalene, anthracene, and acenaphtalene, are linear with a coplanar structure that has no preferred growth direction. This structural feature and the formation of a mesophase mentioned in Sec. 2.3 are the major factors in the easy conversion of the turbostratic structure into well-ordered graphite planes.[13]

Coal-tar pitch is a mixture of these coke-forming aromatic hydrocarbons. As could be expected, it graphitizes readily and the degree of graphitization is a function of the ratio of aromatic to aliphatic hydrogen, known as hydrogen aromaticity (ratio of aromatic-hydrogen atoms to the total number of hydrogen atoms). In coal-tar pitch, this ratio varies between 0.3 and 0.9. The higher the ratio, the more graphitic the coke.[12]

Some aromatic compounds do not form coke but char instead (see Sec. 3.5 below).

3.4   Graphitization Mechanism of Cokes

Graphitization occurs in a series of steps which begins as the increasing temperature passes the carbonization temperature, i.e., ~ 1200°C.[13] Hydrogen, sulfur, and other elements, which might still be present after carbonization, are gradually removed and, as the temperature reaches 2000°C, essentially none remains.[3]

Above 1800°C, the conversion from a turbostratic structure to a graphitic structure (shown in Figs. 3.4 and 3.2 of Ch. 3) begins slowly at first then more rapidly as the temperature passes 2200°C. The gradual graphitization of the structure is readily confirmed by x-ray diffraction (Fig. 4.5).

The crystallite size (L_c) increases from 5 nm, which is a typical size for turbostratic crystallites, to approximately 100 nm or more. At the same time, the interlayer spacing (d) is reduced from 0.344 nm to a minimum of 0.335 nm, which is the spacing of the graphite crystal.

In addition to the increase in L_c and the decrease in d, the graphitization mechanism includes the following steps: (a) removal of most defects within each graphite layer plane as well as between the planes, (b) gradual shifting and growth of the crystallites, (c) removal of cross-linking bonds, (d) evolution of the ABAB stacking sequence, and (e) shifting of carbon rings or single atoms to fill vacancies and eliminate dislocations (such vacancies and dislocations are shown schematically in Figs. 3.5 and 3.6 of Ch. 3).

Graphitization is accompanied by a weight loss, which is attributed to the removal of interlayer chemical species, mostly interstitial carbon.[15] Most graphitizable materials (cokes) require a temperature of 3000°C in order to reach full graphitization with a minimum value of the interlayer spacing, as shown in Fig. 4.6 and 4.7.[15][16] Also shown in Fig. 4.7 is the effect of the duration of graphitization (residence time). At 3000°C, full graphitization is usually obtained within 2 to 3 hours. Lower temperatures require considerably more time.

The process of graphitization can be accelerated by the presence of a metal catalyst or an oxidizing gas. In the latter case, obstacles to the ordering of the graphite layers, such as the more structurally disordered regions and the cross-linking bonds, are preferentially oxidized.

Graphitization can also be enhanced by the use of pressure. At high pressure, a greater degree of graphitization can be achieved at a given temperature than at atmospheric pressure. Likewise, high pressure results in a shorter heat-treatment time or a lower heat-treatment temperature.[2]

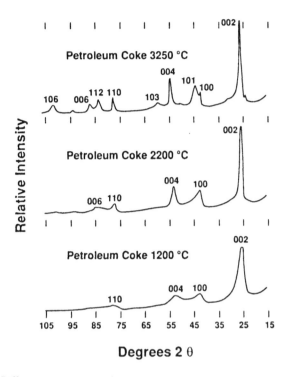

Figure 4.6. Diffraction patterns of petroleum coke as a function of graphitization temperature.[15]

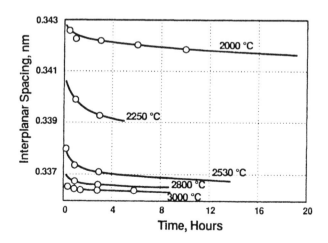

Figure 4.7. Basal plane spacing as a function of time for various temperatures of graphitization.[16]

The kinetics of graphitization of cokes has been shown to be a growth process and not a nucleation-based process with usually a single-valued activation energy, reported at 962 ± 60 kJ/mol (230 ± 15 kcal/mol) in the 2300 - 2900°C temperature range for typical industrial cokes.[17]

3.5 Graphitization of Chars

Graphitization of Aromatics. As mentioned above, not all aromatic hydrocarbons form coke. Some, such as phenanthrene and biphenyl, do not graphitize and are considered char formers. These compounds are branched aromatics (as opposed to the linear structure of the coke-former aromatics) with a preferred axis of growth as shown in Fig. 4.3. This characteristic prevents the formation of extensive graphitic planes and of a liquid mesophase.[2]

Graphitization of Aliphatics. Chars are produced by the carbonization of aliphatic hydrocarbons (compounds with an open-ended chain) and of most polymers. They do not graphitize readily. Their turbostratic structure and the random arrangement of their crystallites (shown in Fig. 3.4 of Ch. 3) remain essentially unchanged, regardless of the temperature and duration of the heat-treatment. However, in some cases the graphitization of chars can be enhanced by a radiation treatment or by the presence of a metallic or mineral catalyst.[18] The catalytic process involves the dissolution of carbon particles at the catalyst sites and the precipitation of graphite.

Graphitization of Polymers. As mentioned above, most polymers, such as those listed in Table 4.1, are char-formers and do not graphitize to any extent (with the notable exception of polyvinyl chloride), although some reduction in the interlayer spacing (d) is usually observed. This d-spacing, however, does not shrink below 0.344 nm for most polymers. This is shown in Fig. 4.8, where the d-spacing of carbonized polyacrylonitrile (PAN) is plotted as a function of graphitization temperature. The crystallite size (L_c) remains small, going from ~ 1.5 to ~ 2.8 nm.[19] Pitch-based fiber, on the other hand, are cokes and graphitize more readily (see Ch. 8).

During carbonization of PAN, an extensive-random network of graphitic ribbons is formed which has a stable configuration. In addition, it is speculated that some tetrahedral (sp^3) (diamond) bonds are formed (which would account for the hardness of these materials).[2] The presence of diamond structure in a similar material, diamond-like carbon, is well established (see Ch. 14). The diamond structure is reviewed in Ch. 2, Sec. 3.0. These two factors, ribbon network and tetrahedral bonds, would prevent further ordering of the structure, regardless of the graphitization temperature.

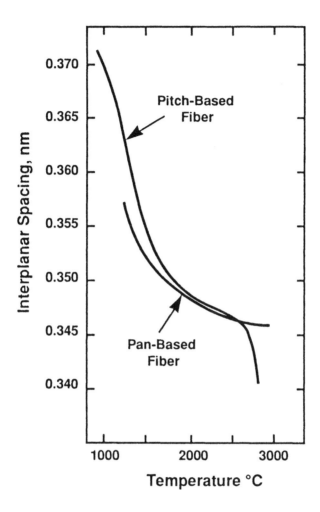

Figure 4.8. Basal plane spacing of PAN-based and pitch-based fibers as a function of graphitization temperature.[8]

Some polymeric films such as polyimide (Kapton), polyacrylonitrile (PAN) and polyfurfural alcohol, when carbonized between sheets of the mineral montmorillonite, form a two-dimensional graphitic structure with highly oriented layer planes which graphitizes readily.[20][21]

REFERENCES

1. *Graphite, A Refractory Material,* Technical Brochure, Carbone Lorraine, Gennevilliers, France (1990)
2. Jenkins, G. M. and Kawamura, K., *Polymeric Carbons*, Cambridge Univ. Press, Cambridge UK (1976)
3. Inagaki, M., et al, *Carbon*, 27(2):253-257 (1989)
4. Fitzer, E., *Carbon*, 25(2):163-190 (1987)
5. Ayache, J., Oberlin, A. and Inagaki, M., *Carbon*, 28(2&3):353-362 (1990)
6. Walker, P. L., *Carbon*, 24(4):379-386 (1986)
7. Kochling, K. H., McEnaney, B., Muller, S. and Fitzer, E., *Carbon*, 23(5):601-603 (1985)
8. Akezuma, M. et al, *Carbon*, 25(4):517-522 (1987)
9. Honda, H., *Carbon*, 26(2):139-136 (1988)
10. Mochida, I., Shimimzu, K., and Korai, Y., *Carbon*, 28(2&3):311-319 (1990)
11. Walker, P. L., Jr., *Carbon*, 28(2&3):261-279 (1990)
12. Eser, S. and Jenkins, R. G., *Carbon*, 27(6):877-887 (1989)
13. Lim, Y. S. and Lee, B. I., Effect of Aromatic Hydrocarbon Addition on Mesophase Formation, *Fiber-Tex 1990*, (J. D. Buckley, ed.), NASA Conf. Publ. 3128 (1991)
14. Eggers, D. F., Jr., Gregory, N. W., Halsey, J. D., Jr. and Rabinovitch, B. S., *Physical Chemistry*, John Wiley & Sons, New York (1964)
15. Mantell, C. L., *Carbon and Graphite Handbook,* Interscience Publishers, New York (1968)
16. Kawamura, K. and Bragg, R. H., *Carbon*, 24(3):301-309 (1986)
17. Murty, H. N., Biederman, D. L. and Heintz, E. A., *Carbon*, 7:667-681 (1969)
18. Mochida, I., Ohtsubo, R., and Takeshita, K., *Carbon*, 18(2&3):25-30 (1990)
19. Cowlard, F. and Lewis. J., *J. of Mat. Science* 2:507-512 (1967)
20. Sonobe, N., Kyotani, T. and Tomita, A., *Carbon*, 29(1):61-67 (1991)
21. Sonobe, N., Kyotani, T., and Tomita, A., *Carbon*, 28(4):483-488 (1990)

5

Molded Graphite: Processing, Properties, and Applications

1.0 GENERAL CONSIDERATIONS

Molded graphite can be defined as a synthetic (or artificial) graphitic product manufactured by a compaction process from a mixture of carbon filler and organic binder which is subsequently carbonized and graphitized. Parts of considerable size, weighing several hundred kilograms, such as the electrodes shown in Fig. 5.1, are manufactured in large quantities.[1]

The basic process was invented by E. G. Acheson, who produced the first molded graphite in 1896. The original applications of molded graphite were electrodes for electric-arc furnaces and movie projectors. Many improvements have been made since then and the applications have increased considerably in scope. Molded graphite is found in almost every corner of the industrial world and forms the base of the traditional graphite industry.

It is often difficult to obtain details of a specific process, particularly if such details are not protected by a patent or cannot be revealed by suitable analyses. Most graphite producers claim that such secrecy is necessary because of the high cost of developing new grades of molded graphite, and the need for the new product to remain ahead of competition long enough for the producer to recover his expenses and realize a profit.[2] Fortunately, a great deal of information on the basic materials and processes is disclosed in the open literature.

Figure 5.1. Graphite electrode. *(Photograph courtesy of Carbon/Graphite Group Inc., St. Marys, PA.)*

2.0 PROCESSING OF MOLDED GRAPHITES

2.1 Raw Materials (Precursors)

Raw Materials Selection. The selection of the appropriate raw (precursor) materials is the first and critical step in the manufacturing process. It determines to a great degree, the properties and the cost of the final product. The characteristics of these raw materials such as the particle size and ash content of cokes, the degree of carbonization of pitch, the particle structure of lampblack, and the impurities and particle size of natural graphite must be taken into account.

To use high-grade, expensive raw materials to produce an undemanding product, such as a grounding anode for electrolytic protection, would be wasteful and economically unsound since these electrodes do not require optimum properties and cost is the overriding factor. On the other hand, nuclear applications demand a graphite with the lowest-possible impurities and the highest-possible mechanical properties. This requires the selection of premium-grade precursor materials with cost somewhat secondary.

Raw materials can be divided into four generic categories: fillers, binders, impregnants, and additives.[1]-[4]

Fillers. The filler is usually selected from carbon materials that graphitize readily. As mentioned in Ch. 4, such materials are generally cokes, also known in industry as "soft fillers". They graphitize rapidly above 2700°C (the graphitization process is described in Sec. 2.4 below). Other major fillers are synthetic graphite from recycled electrodes, natural graphite, and carbon black (see Ch. 10).

Petroleum coke is the filler of choice in most applications. It is a porous by-product of the petroleum industry and an almost-pure solid carbon at room temperature. It is produced by destructive distillation without the addition of hydrogen, either by a continuous process (fluid coking) or, more commonly, by a batch process (delayed coking).

The batch process consists of heating high-boiling petroleum feedstocks under pressure to approximately 430°C, usually for several days.[5] This promotes the growth of mesophase-liquid polycylic crystals. The material is then calcined up to 1200°C, to remove almost all the residual hydrogen, and finally ground and sized.

By varying the source of oil and the process parameters, it is possible to obtain various grades of petroleum-coke filler with different properties. The industry commonly uses three grades:

- Needle coke, a premium grade with distinctive needle-shape particles, produced by delayed coking from selected feedstocks with low concentration of insolubles. It is used in applications requiring high thermal-shock resistance and low electrical resistivity.

- Anode coke for less demanding applications.

- Isotropic coke in applications where isotropic properties and a fine-grained structure are required.

Binders. The most common binder is coal-tar pitch which is a hard, brittle and glassy material, described in Ch. 4, Sec. 2.3. It is a by-product of metallurgical-coke production and is obtained by the distillation or heat treatment of coal-tar. From 35 to 60 kg of pitch are produced from every metric ton of coal.

The composition of coal-tar pitch is complex and may vary considerably since it depends on the degree of refinement of the available coke-oven tars. Two factors can noticeably influence the quality and graphitization characteristics of the pitch: *(a)* its softening point and *(b)* the content of insoluble complexes of quinoline (C_9H_7N). This content may vary widely from one pitch to another.[6]

Other binders such as petroleum pitch and thermosetting resins are used for specialty applications.

2.2 Production Process

A typical production-process flow diagram for molded graphite is shown in Fig. 5.2.[1][2] The production steps are as follows.

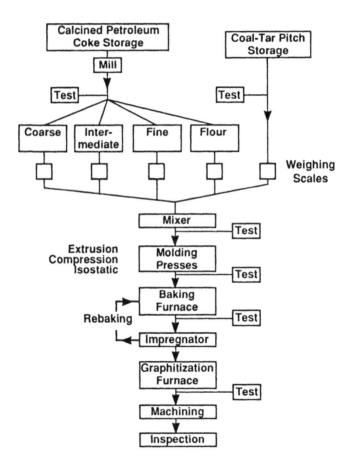

Figure 5.2. Production-process flow diagram of molded graphite.[1][2]

Milling and Sizing. Filler and binder are ground or milled to the particle-size requirement which may vary from 1 μm (flour) to 1.25 cm. A batch usually consists of more than one size. This allows better control of the packing characteristics and optimizes the density of the final product. Table 5.1 lists the grain size of various grades of molded graphites and its effect on properties.[3]

Table 5.1. Particle Sizes and Characteristics of Graphite Grades

Grade	Grain Size	Properties
Medium grain	Up to 1.25 cm	Low density Low thermal expansion Low strength High permeability
Fine grain	0.05 to 0.15 cm	Medium density Medium thermal expansion Medium strength Medium permeability
Micrograin	<1 μm to 75 μm	High density High thermal expansion High strength Low permeability

Mixing. Filler and binder are weighed in the proper proportion and blended with large mixers into a homogeneous mix where each filler particle is coated with the binder. Blending is usually carried out at 160 - 170°C, although temperatures may reach as high as 315°C on occasion. When mixing at lower temperatures (below the melting point of the binder), volatile solvents such as acetone or alcohol are often added to promote binder dispersion.

The final properties of the molded product are controlled to a great degree by the characteristics of the filler-binder paste such as: *(a)* the temperature dependence of the viscosity, *(b)* the general rheological behavior, and *(c)* the hydrodynamic interaction between filler particles.[7]

Forming Techniques. Three major techniques are used to form the graphite mix: extrusion, compression (uniaxial loading), and isostatic pressing. They are shown graphically in Fig. 5.3.

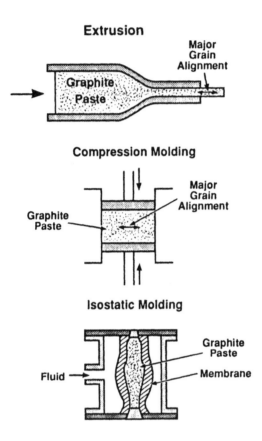

Figure 5.3. Forming techniques for molded graphites.

The selection of a given technique has a great influence on the final properties of the molded product as shown in Table 5.2.

Extrusion. Extrusion is a major technique which is favored for the production of parts having a constant cross-section, such as electrodes. The mix is cooled to just above the softening point (approximately 125°C), then extruded through steel dies, cut to length and rapidly cooled to solidify the pitch before distortion occurs. The resulting shape is known in the industry as a "green shape".

Table 5.2. Characteristics of Forming Techniques

Technique	Characteristics
Extrusion	Anisotropic properties Non-uniformity of cross-section Presence of flow lines and laminations Limited to parts of constant cross-section Production of large parts possible Low cost
Compression	Non uniformity Edge effect Presence of flow lines and laminations Medium cost
Isostatic	Isotropic properties Uniformity No flow lines or laminations High cost

Extrusion pressures are on the order of 7 MPa (100 psi). Some alignment of the coke-filler particles takes place which imparts anisotropy to the properties of the finished product. This anisotropy can be controlled to some extent by changing the mix formulation and the extrusion geometry.[4] The center of the extruded material is usually of lower quality than the material near the outside edge and defects such as flow lines and laminations are difficult to avoid. On the plus side, it is the lowest-cost technique which is satisfactory for most large parts, such as furnace electrodes. It represents the largest tonnage of molded graphite.

Compression (Uniaxial) Molding. The mix in compression molding is usually a fine powder (flour) as opposed to the coarser material used in extrusion. Tungsten carbide dies are frequently used with pressures on the order of 28 to 280 MPa (4000 to 40,000 psi). Complex shapes can be produced by this process (Fig. 5.4).[2] However, die-wall friction and die edge effect may cause non-uniformity in the density and other properties of the finished product.

Isostatic molding. In isostatic molding, pressure is applied from all directions through a rubber membrane in a liquid-filled chamber, resulting

Figure 5.4. Examples of molded graphite shapes. *(Photograph courtesy of Pure Industries, St. Marys, PA.)*

in a material with great uniformity, isotropic properties, and generally with few defects. However, the molding process is expensive and cost is higher than extrusion or compression molding.

2.3   Carbonization, Graphitization, and Machining

Carbonization. Carbonizing (also known as baking) the green shape is the next step (see Ch. 4, Sec. 2). Carbonization takes place in a furnace in an inert or reducing atmosphere. The process may last from a few days to several weeks depending on the constituents, and the size and geometry of the part. The temperature is raised slowly to 600°C, at which stage the binder softens, volatiles are released and the material begins to shrink and harden. Typical shrinkage is 6%. The parts must be supported by a packing material to prevent sagging.

The temperature is then raised to 760 to 980°C (or up to 1200°C in special cases). This can be done faster than the first temperature step, since most of the volatiles have by now been removed, the material is already hard, and sagging is no longer a problem.

Impregnation. After the carbonization stage, the material has a high degree of porosity. To further densify it, it is necessary to impregnate it with coal-tar pitch or a polymer such as phenolic. Impregnation is usually carried out in a high-pressure autoclave and the carbonization process is repeated. In special, limited-use applications, non-carbon impregnating materials such as silver and lithium fluoride impart specific characteristics, particularly increased electrical conductivity.[8]

Graphitization. During graphitization, the parts are heated up to 3000°C (see Ch. 4, Sec. 3). The temperature cycle is shorter than the carbonization cycle and varies depending on the size of the parts, lasting from as short as a few hours to as long as three weeks. It is usually performed in a resistance furnace (the original Acheson cycle) or in a medium-frequency induction furnace.

Graphitization increases the resistance of the material to thermal shock and chemical attack. It also increases its thermal and electrical conductivities.

Puffing. Puffing is an irreversible expansion of molded graphite which occurs during graphitization when volatile species, such as sulfur from the coke, are released. Puffing is detrimental as it causes cracks and other structural defects. It can be eliminated (or at least considerably reduced) by proper isothermal heating and by the addition of metals or metal compounds with a high affinity for sulfur.[9]

Purification. For those applications that require high purity such as semiconductor components and some nuclear graphites, the material is heat-treated in a halogen atmosphere. This treatment can remove impurities such as aluminum, boron, calcium, iron, silicon, vanadium and titanium to less than 0.5 ppm.[10] The halogen reacts with the metal to form a volatile halide which diffuses out of the graphite. The duration of the treatment increases with increasing cross section of the graphite part.

Machining. The graphitized material can now be machined to the final shape. Since it is essentially all graphite, machining is relatively easy and is best performed dry to avoid water contamination. Common cutting tool materials are tungsten carbide, ceramic and diamond. A good ventilation system is necessary to control and collect powdery dust.[11]

The details of material formulation, processing steps, and equipment in the graphite-molding process can vary considerably from one manufacturer to another. In many cases, these details are jealously guarded and most companies consider their proprietary processes essential to their economic survival.

3.0 CHARACTERISTICS AND PROPERTIES OF MOLDED GRAPHITE

3.1 Test Procedures and Standards

Molded graphites are usually characterized by the following properties:

- Apparent density
- Electrical resistivity
- *Young's modulus*
- Coefficient of thermal expansion (CTE)
- Flexural strength

The test methods for these and other useful properties are listed in Table 5.3. When appropriate, the tests are performed both "with the grain" and "perpendicular to (across) the grain."

The hardness is generally measured with a Shore model C-2 Scleroscope with a specially calibrated diamond-tip hammer.[2]

Table 5.3. Test Methods for Molded Graphite

Hardness (Scleroscope)	ANSI/NEMA CB1-1977 ASTM C-886
Density, Porosity	ANSI/NEMA CB1-1977 ANSI C-64.1 ASTM C559-79 ASTM C 2080a ASTM C819-77
Electrical Resistivity	ANSI/NEMA CB1-1977 ASTM C611-69(76)
Flexural Strength	ANSI/NEMA CB1-1977 ANSI C-64.1 ASTM C651-70(77)
Compressive Strength	ASTM C695-75
Tensile Strength	ASTM C749-73(79)
Thermal Expansion	ASTM E228-71(79)
Modulus of Elasticity	ASTM E111-61(78)
Ash Content	ANSI/NEMA CB1-1977 ASTM C4561-69(79)
Moisture Content	ASTM C562-69(79)

Testing for flexural strength is critical and the results can vary widely depending on the sample size and geometry and on the test method.[1] For instance, the flexural strength of a square cross-section of a molded graphite, tested in a four-point test fixture, can be 40% lower than the measured strength of the same material with a circular cross-section tested in a three-point loading fixture. For that reason, comparing the strength of various molded graphites can be meaningless, if the test methods are different.

The thermal coefficient of expansion (CTE) of graphite is not a fixed value but varies with temperature. For that reason, the CTE should always be quoted with the range of temperature of the test. When testing above 350°C, it is necessary to do so in an inert atmosphere to avoid oxidation of the graphite.

The properties of molded graphite are reviewed in the following sections. The values given are taken from the open literature and manufacturer's data sheets and must be viewed accordingly.

3.2 Density and Porosity

The real density of molded graphite, that is the density exclusive of open porosity (measured with a pycnometer) but including closed-internal porosity, is 2.20 - 2.23 g/cm^3. This value is near the theoretical density of the graphite crystal (2.25 g/cm^3), indicating that the amount of internal porosity is small and the structure of the material is well ordered.

The apparent density of molded graphite, that is the density inclusive of porosity, varies considerably, depending on the process, the impregnation cycle and other factors. Values generally quoted in the industry are 1.4 - 2.0 g/cm^3, corresponding to an open porosity of 10 - 35%. A common method of measuring apparent density is the Avogadro (or water-weight) method. To avoid overstating the result, air bubbles must be eliminated and the part must be perfectly dry before testing.

Typical pore-size distribution of several molded graphite materials is shown in Fig. 5.5.[12]

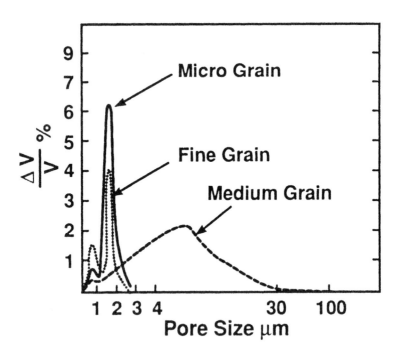

Figure 5.5. Pore-size distribution of various grades of molded graphite.[12]

3.3   Effect of Particle (Grain) Size on Properties

The size of the particle (grain size), can have a considerable effect on the properties of molded graphite as shown in Table 5.4.

Table 5.4. Effect of Particle Size on Properties [10]

Grade**	Particle Size mm	Density g/cm^3	Specific Resistance 0.cm	Flexural Strength MPa	CTE* m/m$^\circ$C x 10^{-6} (100 - 600°C)
AI1RL	0.075	1.72	0.0016	28	3.2
AI4RL	0.025	1.82	0.0018	42	5.7
AI5RL	0.010	1.90	0.0020	77	7.0

Notes: * CTE= coefficient of thermal expansion

**Materials are isostatically pressed grades from The Carbon/Graphite Group Inc., St. Marys, PA

As shown in the above table, reducing particle size (grain size) results in an substantial increase in the apparent density and, correspondingly, a lowering of the porosity. The flexural strength is almost tripled. On the negative side, the CTE is more than doubled, which may lead to a decrease in thermal shock resistance. The effect on the electrical resistivity is not as pronounced.

Particle size also affects the finishing since, during machining operations, particles can be torn out of the surface by the tool, leaving a void and the larger the particle, the larger the void. When a smooth finish is required with close tolerances, a small particle size material must be selected.

3.4   Effect of Grain Orientation on Properties

The degree of grain orientation varies with the production process as mentioned above and has a considerable effect on the properties of molded graphite as shown in Table 5.5.

Table 5.5. Effect of Grain Orientation[13]

Grade	Particle Size mm	Density g/cm³	Specific Resistance 0.cm	Flexural Strength MPa	Compress Strength MPa	CTE* (m/m °Cx10⁻⁶) (100-600°C)
6250 extruded	0.8	1.67	0.0007 WG 0.0012 AG	20 WG 12 AG	45 WG 27 AG	2.3 WG 3.8 AG
2020 isostatically pressed	0.04	1.77	0.0018 WG 0.0018 AG	38 WG 33 AG	89.6 WG 65.5 AG	3.2 WG 3.8 AG

Notes: * CTE = coefficient of thermal expansion
WG = with grain
AG = across grain
Materials and data from Carbone of America, Graphite Materials Div., St. Marys, PA

The extrusion process tends to align the particles in the extrusion flow direction and extruded graphite has a pronounced anisotropy, particularly in the strength and electrical properties. The isostatic-pressing process, on the other hand, results in a material that is much more uniform and is generally considerably stronger.

3.5 Mechanical Properties

Modulus and Strength. The mechanical behavior of molded graphite is essentially that of glasses or ceramic materials, i.e., the material is brittle and its strength is a function of the bond strength between crystallites. At ambient temperature, these bonds resist plastic deformation and mechanical failure of molded graphite occurs by brittle fracture. Porosity, flaws, and structural defects concentrate the applied stresses and the material fails at much-lower levels than the theoretical strength of the graphite crystal.

As the temperature is raised, the flaws and structural defects are gradually annealed, stresses are relieved, and plastic deformation becomes possible. The result is a general increase in strength and modulus with increasing temperature. At 2500°C, the strength is almost twice the room-temperature strength. Above 2500°C, plastic deformation increases and strength decreases. Elongation, which is negligible at ambient temperature, becomes appreciable then and reaches 20% at 2800°C. The effect of temperature on strength and modulus is shown in Fig. 5.6.

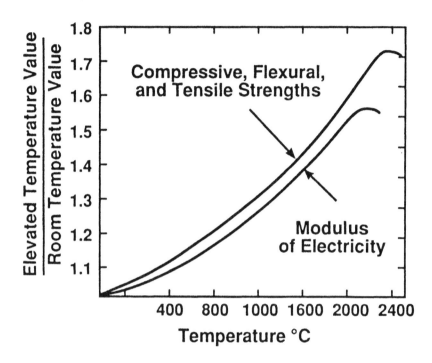

Figure 5.6. Effect of temperature on strength and modulus of molded graphite.[13]

Since molded graphite does not deform plastically, at least at low temperatures, measurement of its tensile strength is difficult and unreliable. Instead, it is preferable to measure flexural (transverse) strength, which is a more reproducible property. Tensile strength is generally 50 to 60% of the flexural strength and compressive strength is approximately twice as much.

Molded graphite has lower strength than metals and most ceramics at ambient temperature. However, at high temperatures this is no longer the case and, above 2000°C, molded graphite (and other graphite products) is a superior structural material. This is shown in Fig. 5.7, where the specific strength of graphite and other high-temperature materials are shown. Since molded graphite has low density compared to these materials, its specific strength is relatively high.

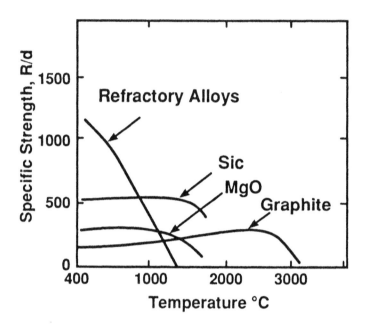

Figure 5.7. Specific strength as a function of temperature of graphite and other high-temperature materials.[12]

The room-temperature transverse (flexural) strength of molded-graphite grades varies between 10 and 100 MPa and its Young's modulus (also at room temperature) is between 5 and 10 GPa. These large variations are due to differences in raw materials and processing techniques. Flexural testing is described in detail in the test procedures listed in Table 5.3. In the more common three-point loading test, typical test specimens have a square cross section, 5.08 mm (0.200 in) or 3.18 mm (0.125 in) on the side. The support span is 3.05 mm (1.20 in) or 1.52 mm (0.60 in) for the larger and smaller test specimens, respectively. The strength is given by the following formula:

$$S = 3PL/2bd^2$$

where: S = flexural strength
           P = load at rupture
           L = support span
           b = specimen width
           d = specimen thickness[2]

Hardness. As mentioned in Ch. 3, the distance between the basal planes of the graphite crystal is relatively large (0.334 nm) and the *van der Waals bond* between these planes is weak (7.1 kJ mol). The interplanar distance is further increased by lattice defects, interstitial foreign atoms and other irregularities. As a result the basal planes can slip easily over each other without loosing their coherence (as long as the slip is not hindered by cross-linking boundary conditions), and graphite is able to yield plastically.

This characteristic makes it difficult to interpret hardness measurements except with the ball-identation method which can be considered reasonably accurate. The resulting "contact hardness" is defined as the average pressure required to indent the material to a depth equal to 2/100th. of the radius of the ball. Other hardness-measurement methods such as the Scleroscope (which is the measure of the rebound height of a falling diamond-tipped hammer) are convenient but, as they are not based on the same principle, cannot readily be correlated with the ball hardness.

Table 5.6 lists typical contact- and Scleroscope-hardness values of several types of molded graphite.[13]

Table 5.6. Hardness of Molded Carbon and Graphite

Material	Contact Hardness kg/mm^2	Scleroscope Hardness
Electrographite	18	40 - 80
Hard carbon	35 - 50	70 - 100
Coke-based carbon		70 - 90
Lampblack-based carbon		70 - 110

Frictional Properties. Molded graphite materials have inherently low friction due to the ease of basal-plane slippage and the resulting low shear strength mentioned above.

When graphite is rubbed against a metal or ceramic surface, a thin transfer film is formed on the rubbed surface. This lowers the coefficient of friction which can be less than 0.01 after the transfer film is fully developed.

This hydrodynamic film is maintained only in the presence of adsorbable vapors such as water vapor, oxygen or contaminating organic gases. In a chemically pure atmosphere of inert gases or nitrogen, in a high vacuum, or at high temperature, the film is no longer formed or, if formed, is readily broken. As a result, the coefficient of friction becomes high (~ 1.0) and wear becomes excessive, up to five orders of magnitude greater than in the vapor environment. This phenomenon is known as "dusting". Under these conditions, graphite is not an effective lubricant and lubricity additives and intercalation compounds are necessary (see Ch. 10).

3.6   Thermal Properties

Thermal Conductivity: The thermal conductivity of the ideal graphite crystal was reviewed in Ch. 3, Sec. 4.3. The mechanism of heat transfer is by lattice vibration and the thermal conductivity is approximately 200 times greater in the basal plane (*ab* directions) than across the planes (*c* direction), thus reflecting the anisotropy of the graphite crystal.

This anisotropy is less pronounced in molded graphites and carbon materials and the thermal conductivity is more isotropic. Typical values for graphite and other high-conductivity materials are given in Table 5.7.[13]

Table 5.7. Thermal Conductivity of Molded Carbon, Graphite and High-Conductivity Metals

	W/m·K at 20°C
Electrographite (from petroleum coke)	159
Electrographite (from lampblack)	31.4
Carbon	8
Diamond	1800
Silver	420
Copper	385

The relationship between the thermal conductivity of molded graphite and temperature is similar to that of single-crystal graphite. After an initial increase (due to increase in specific heat), the conductivity decreases (due to increasing phonon scattering) and, at 2000°C, it is an order of magnitude lower than at room temperature. This relationship is shown in Fig. 5.8.

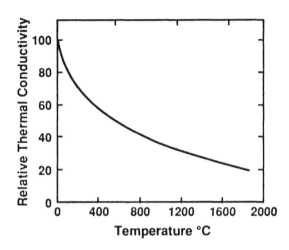

Figure 5.8. Relative thermal conductivity of molded graphite as a function of temperature.[12]

Thermal Expansion. The thermal expansion of the graphite crystal was reviewed in Ch. 3, Sec. 4.4. This expansion has a marked anisotropy. It is low in the *ab* directions (lower than most materials) but an order of magnitude higher in the *c* direction.

Molded-graphite materials have a much more isotropic thermal expansion than the graphite crystal, since the crystallites are randomly oriented and the inherent porosity absorbs a portion of the expansion. Still some degree of anisotropoy remains in extruded- and compression-molded graphites and the coefficient of thermal expansion (CTE) across the grain is typically 50% higher than with the grain. The CTE of isostatically molded graphites can be essentially isotropic.

The CTE of molded graphite increases with temperature, as shown in Fig. 5.9. At 1500°C, it is approximately twice the room temperature value. Yet, it remains low compared to most other structural materials as shown in Table 5.8.[14]

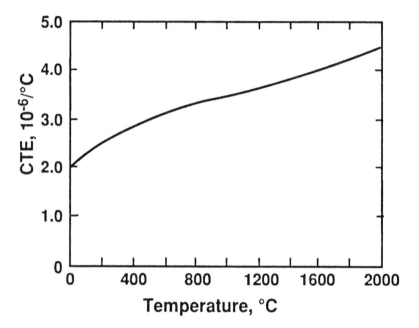

Figure 5.9. Average coefficient of thermal expansion (CTE) of graphite as a function of temperature.

Table 5.8. Thermal Expansion of Typical Molded Graphite and Other Materials

Material	CTE m/m·°C x 10^{-6} (100 - 600°C)
Extruded Graphite	
with grain	1.8 - 2.9
across grain	3.2 - 5.0
Isostatic Graphite	3.2-5.7
Aluminum	23.5 (@ 25°C)
Copper	16.6 (@ 25°C)
Tungsten	4.5 (@ 25°C)

The thermal expansion of molded graphite is an important property since many applications involve high temperature, and expansion values must be known accurately.

Thermal Shock. The low modulus, high thermal conductivity, and low thermal expansion combine to give molded graphite excellent thermal-shock resistance. It is difficult to rupture the material by thermal shock alone.

3.7 Electrical Resistivity

The electrical resistivity of the graphite crystal was reviewed in Ch. 3, Sec. 5. The crystal has a high degree of electrical anisotropy with resistivity that is low in the *ab* directions and high in the *c* direction. Molded graphites (even the isostatically molded material where the crystallites are essentially random) have higher resistivity than the single-crystal in the *ab* directions, but can still be considered electrical conductors.

Table 5.9 lists typical electrical-resistivity values of several molded carbons and graphites and selected low-resistivity metals.

Table 5.9. Electrical Resistivity of Molded Graphite and Selected Metals

Material	Electrical Resistivity at $25^\circ C$, $\mu\Omega$-m
Molded carbon	~ 50
Electrographite (from petroleum coke)	7.6
Electrographite (from lampblack)	30.5
Pyrolytic graphite (*ab* direction)	2.5 - 5.0
Aluminum	0.026
Copper	0.017
Tungsten	0.056
Silver	0.016

The values listed above for the molded carbons and graphites are averages and may vary depending on the raw materials, the process, and the degree of graphitization.

The electrical resistivity of molded graphite, like that of the graphite crystal and metals, increases with temperature, above approximately 400°C, and the material has the positive *temperature coefficient of resistance* which is typical of metals. Below 400°C, the coefficient is usually slightly negative. At high temperature, graphite becomes a better conductor than the most-conductive refractory metal. The effect of temperature on the relative resistivity of molded graphite is shown graphically in Fig. 5.10.[3]

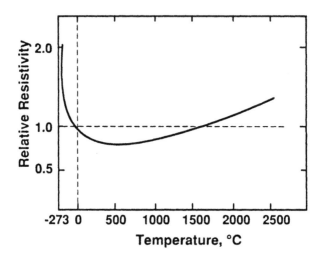

Figure 5.10. Relative variation of electrical resistivity of molded graphite as a function of remperature.[12]

3.8 Emissivity

Emissivity is the ratio of the energy radiated by a body to that radiated by an equal area of a perfect black body. According to the *Stefan-Boltzmann law*, a perfect black body is an ideal material which radiates the maximum amount of energy. The emissivity of a material depends on its structure and on its surface conditions.

Molded graphite, being black, has high emissivity, which is an important advantage in high-temperature applications. Table 5.10 lists the total emissivity of several carbon and graphite materials and selected metals for comparison.[13]

Table 5.10. Emissivity of Carbon and Graphite Materials and Selected Metals

Material	Total Emissivity
Graphite (petroleum-coke base)	0.70 - 0.90
Graphite (lampblack base)	0.85 - 0.95
Molded Carbon	0.60 - 0.80
Lampblack	0.90 - 0.99
Silver	0.04
Nickel, oxidized	0.87
Tungsten, polished	0.15

4.0 APPLICATIONS AND MARKET OF MOLDED GRAPHITE

4.1 General Considerations

The technology of molded graphite is versatile, improvements in the manufacturing techniques are continuously made and the scope of applications is gradually expanding. This expansion is the direct result of a sizeable research effort carried out by many workers in universities, government laboratories, and industry.

In the last twenty years or so, several major applications of molded graphite have developed into an important, complex, and diverse industrial market. Because of this diversity, the classification of these applications is a critical factor if a proper perspective of the industry is to be obtained.

The classification by product function is the primary one used in this book. It is used in this chapter for molded graphite and in the following chapters for the other main types of carbon and graphite materials, i.e., polymeric, pyrolytic, fibers, carbon-carbon and powders. It is based on the following major functions: electrical, structural and mechanical, chemical and refractory, and nuclear. These functions corresponds roughly to the various segments of industry and end-use market, such as electronic and semiconductor, metal processing, chemicals, automobile, and aerospace.

The world market for molded graphite is large as shown in Table 5.11.[1]

Table 5.11. Estimated World Market for Molded Graphite in 1991

	($ millions)
Steel Production Electrodes	2250
Aluminum Production Anodes	940
Cathodes	250
Specialty Graphites	300
Total	3740

The specialty graphites listed in the table include the molded graphites produced for semiconductor, electrical, chemical, nuclear, biomedical, mechanical, and aerospace industries.

4.2   Applications in the Metal Processing Industry

Electrodes for the Production of Electric-Arc Steel. As shown in Table 5.11, the production of electrodes for steel and aluminum processing is the largest application of molded graphite, in terms of both tonnage and dollars. As mentioned in Sec. 1.1, electrodes are one of the original applications and have been manufactured with essentially the same process for almost a century.

The largest use of these electrodes is in the production of steel in electric-arc furnaces, mostly for the reclamation of ferrous scrap.[1] These electrodes must have good electrical conductivity, good refractory properties, and low cost. They gradually erode in use and at times break altogether and must be replaced at regular intervals.

The worldwide production of steel is slowly increasing, but the production of electric-arc furnace steel is increasing at a more rapid rate. This rate reached 28% of total production in 1990 as shown in Fig. 5.11. The major producers by area are Europe, North America, and Japan.

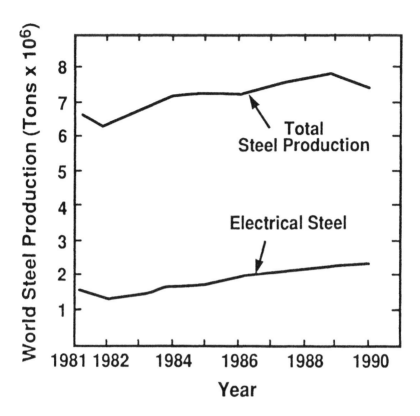

Figure 5.11. World production of steel.[1]

The gradual increase in electric-arc steel production does not necessarily translate into a gradual increase in the tonnage of graphite electrodes. In fact the opposite is happening and the total consumption of graphite electrodes is decreasing (Fig. 5.12) due to a pronounced decrease in the consumption of graphite per ton of electric steel produced. In 1975, an average of 7.5 kg of molded graphite was required to produce one ton of electric-arc steel. In 1990, this amount had dropped to 5 kg. The downward trend is continuing.

This reduction is the result of two factors: *(a)* improvements of the properties of the graphite materials, particularly a considerable reduction of the thermal expansion and resulting increase in thermal-shock resistance, and *(b)* improvements in arc-furnace operating techniques.

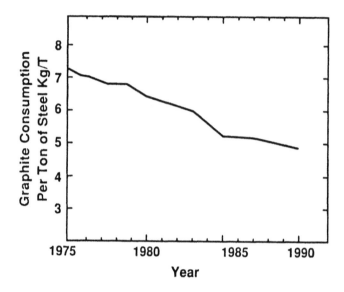

Figure 5.12. Consumption of graphite in the production of steel.[1]

Electrodes for Aluminum Production. Aluminum is processed electrolytically and the production of the necessary electrodes is the second-largest application of molded graphite (see Table 5.11 above).[1] The anodes are similar to those used in electric-arc steel production and are also manufactured from petroleum-coke filler and coal-tar pitch. The aluminum collects at the cathodes which are large blocks lining the electrolytic cell. These cathodes were originally made of baked carbon based on anthracite coal but, in recent years, have been upgraded and are now made of molded graphite from petroleum coke.

The world production of aluminum was estimated at 14.2×10^6 metric tons in 1990. The consumption of molded-graphite anode is approximately 400 kg per metric ton of aluminum and the total estimated consumption worldwide is 5.68×10^6 metric tons.

Melting, Smelting, and Casting of Metals. Molded graphite has numerous applications in the processing of ferrous and non-ferrous metals and alloys such as copper, copper-nickel, brass, bronze, zinc, aluminum alloys, nickel and its alloys, precious metals, and grey and ductile irons.[3][15] The wide variety of these applications is shown in the following partial list:

- Molds for centrifugal casting of brass and bronze bushings and sleeves
- Molds for pressure casting of steel slab and railroad-car wheels
- Dies for continuous casting and extrusion of aluminum and other non-ferrous metals
- Extrusion guides and run-out tables
- Hot-pressing molds and plungers
- Pumps for molten aluminum and zinc
- Brazing fixtures
- Furnace linings
- Sintering boats and trays

The factors to consider in selecting a suitable grade of molded grahite for metal processing are: high thermal conductivity to reduce the thermal stresses; high hardness to reduce abrasive wear and improve the lifetime of the die; and high density, fine grain, and low porosity to minimize chemical attack by molten metals and by dissolved atomic oxygen (the latter especially found in nickel and its alloys). Fig. 5.13 shows typical molded-graphite casting dies.

Figure 5.13. Examples of graphite casting dies. *(Photograph courtesy of Sigri Great Lakes Carbon Corp., Niagara Falls, NY.)*

4.3   Semiconductor and Related Applications

In 1947, Bardeen, Brattain, and Shockley of Bell Telephone Laboratories demonstrated the transistor function with alloyed germanium and this date is generally recognized as the start of the solid-state semiconductor industry. The era of integrated circuits (IC's) was inaugurated in 1959, when, for the first time, several components were placed on a single chip at Texas Instruments.

These developments resulted in a drastic price reduction in all aspects of solid-state circuitry and the cost per unit of information (bit) has dropped by an estimated three orders of magnitude in the last twenty years. This cost reduction been accompanied by a similar decrease in size, and today circuit integration has reached the point where more than a million components can be put on a single chip.

This progress is largely due to the development of glass and ceramic fabrication techniques, single-crystal production processes, and thin-film technologies such as evaporation, sputtering and chemical vapor deposition (CVD). These advances were made possible in part by the availability of high-purity molded graphite and its extensive use as molds, crucibles and other components as shown by the following examples.

Molded Graphite for Crystal Pulling. Single crystals of silicon, germanium and *III-V and II-VI semiconductors* are usually produced by the ribbon or the Czochralski crystal-pulling techniques. The latter is shown schematically in Fig. 5.14. The process makes extensive use of molded graphite, as shown on the figure. The crucible holding the molten material is made of high-purity graphite lined with quartz, and so are the support and the heater. In some cases, the crucible is coated with pyrolytic boron nitride deposited by chemical vapor deposition (CVD).

Other Molded Graphite Applications in Semiconductor Processing. The following is a partial list of current applications of molded graphite in semiconductor processing:[16]

- Boats and assemblies for liquid-phase epitaxy
- Crucibles for molecular-beam epitaxy
- Susceptors for metallo-organic CVD
- Wafer trays for plasma-enhanced CVD
- Shields, electrodes, and ion sources for ion implantation
- Electrodes for plasma etching

- Barrel-type wafer holders for epitaxial deposition (Fig. 5.15)
- Liners for electron-beam evaporation
- Rresistance-heated jigs for brazing and glass-to-metal sealing
- Electrodes for polycrystalline-silicon deposition
- Boats for reduction heating of germanium oxide
- Anodes for power tubes and high-voltage rectifiers

In some of these applications, it is necessary to coat the surface of the molded graphite with a more inert coating such as pyrolytic graphite, boron nitride (BN) or silicon carbide (SiC), to prevent contamination and reaction with the graphite at high temperature. The coating is usually done by CVD as reviewed in Ch. 7.

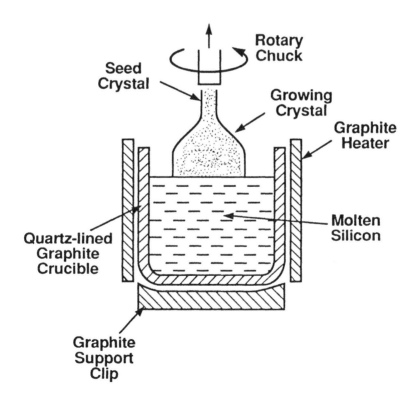

Figure 5.14. Czochralski apparatus for crystal growth of silicon.[3]

Figure 5.15. Graphite barrel holder for semiconductor wafer processing. *(Photograph courtesy Sigri Great Lakes Corp., Niagara Falls, NY.)*

4.4 Electrical Applications

Electrical applications of molded graphite are well established and the material has been a standard for many years due to its chemical inertness, good electrical conductivity and ability to withstand the heat of electrical arcing with minimum damage. The following is a list of current applications:

- Brushes for electric motors
- Current-collecting shoes for railroad conductor rails
- Heating elements for high-temperature furnaces
- Electrodes for lighting, electrical-discharge machining (EDM), and electric-arc furnaces
- Fuel-cell components
- Anodes, cathodes, and buss bars of zinc-chloride, zinc-bromide, and other advanced batteries

4.5   Mechanical Applications

Fine-grain, high-density, and low-porosity molded graphite is extensively used for bearings and seals, particularly in high-temperature applications (up to 600°C) where conventional liquid lubricants degrade rapidly. It has excellent lubricating and frictional properties (reviewed in Sec. 2.5 above), it is chemically resistant, its high thermal conductivity helps dissipate the heat generated by rubbing action and, at high temperature (>500°C), its compressive strength is superior to that of most other engineering materials.

However molded graphite cannot be used to seal materials that would attack it chemically, such as aqua regia, perchloric acid, and oleum (fuming sulfuric acid) (see Ch. 3, Sec. 7.0). Sealing failure may also occur with abrasive materials such as chromic acid, chromic oxide, chrome plating compounds, potassium dichromate, sodium chromate, and sodium dichromate.[2]

The following is a partial list of current commercial applications of bearings and seals:

- Seal rings for gas-turbine engines to protect compressor-shaft bearing and engine oil system for 600°C gases, at rubbing speed of 150 m/s and 10,000 hours service.

- Seal rings in chemical pumps for corrosive-fluid transfer with a rubbing speed of 15 m/s, operating up to 100°C for 10,000 hours.

- Seal rings for water pumps of industrial and truck diesel engine, operating at 120°C and 0.18 N/mm² (25 psi) for 20,000 hours.

- Seals rings for automobile freon air compressor for air conditioning, operating up to 1.75 N/mm² (250 psi) and 93°C.

- Seal rings for home washing machine and dishwasher water pumps, operating at 0.28 N/mm² and to 66°C for 10 years

- O-rings for cylinder heads of race-car engines

Other mechanical applications of molded graphite include glass-lehr roll bearings, roller bearings for jet-engine exhaust nozzles, metering seats for gasoline pumps, vane pumps for air compressors, rupture disks, and many others.

A specialized group of applications is found in aerospace systems which include rocket nozzles and reentry nose cones (shown in Fig. 5.16) where the performance of molded graphite has been excellent, due to its high-temperature strength and resistance to erosion and thermal-shock (see Ch. 9).

Figure 5.16. Graphite reentry nose cone. *(Photograph courtesy Sigri Great Lakes Corp., Niagara Falls, NY.)*

4.6   Chemical Applications

In addition to the seal applications mentioned above, molded graphite has many applications in areas where chemical resistance is the major factor. Such applications are found in chemical reactors, heat exchangers, steam jets, chemical-vapor deposition equipment, and cathodic-protection anodes for pipelines, oil rigs, DC-power lines, and highway and building construction.

4.7 Nuclear Applications

Molded graphite is one of the best material for nuclear-fission applications since it combines high neutron-moderating efficiency and a low neutron-absorption cross section, good mechanical strength and chemical resistance, ease of machinability, and relatively low cost.[17]

However, nuclear radiation affects the crystal lattice which becomes distorted by collisions with fast neutrons, other energetic particles, and displaced carbon atoms. As a result, the properties are altered to some extent. Strength and hardness generally increase and dimensional changes become noticeable particularly at high temperature, as shown in Fig. 5.17.[18][19]

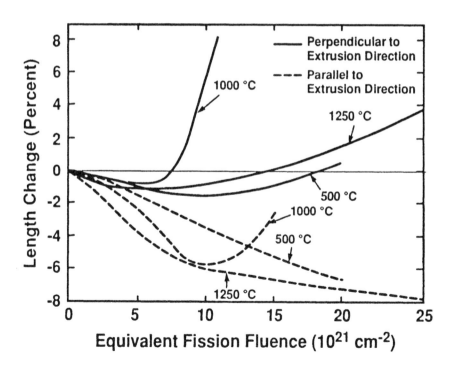

Figure 5.17. Length changes of typical nuclear graphite as a function of fission fluence.[18][19]

Nuclear graphite is usually manufactured from high-grade coke filler and pitch binder, processed as described earlier in this chapter, and then thoroughly purified by halogen treatment (see Sec. 2.3 above).[17] The highest degree of purity is necessary to attain the required nuclear properties.

The material was the building block of CP-1, the world first nuclear reactor at the University of Chicago, and will form the nuclear core for the new generation of high-temperature gas-cooled reactor (HTGR), tentatively scheduled for construction in the mid-90's.[1] It is presently used as fuel-element blocks, replaceable or permanent reflectors, and other components.

Molded graphite is also used in experimental fusion reactors such as the Tokomak Fusion Test Reactor as interior liners, movable limiters, and specialized fixtures, where its low atomic number is an important factor in reducing interference with the fusion reaction.

REFERENCES

1. Gazda, I. W., *Twentieth Biennial Conf. on Carbon*, Univ. of Cal., Santa Barbara, CA (June 1991)
2. Massaro, A. J., *Primary and Mating Ring Materials,* Publ. of Pure Industries, St. Marys, PA (Nov. 1987)
3. *Specialty Graphite,* Technical Brochure, Great Lakes Carbon Corp., St. Marys, PA (1990)
4. Schroth, P. and Gazda, I. W., in *Electric Furnace Steelmaking* (C. R. Taylor, Ed.), 71-79, The Iron and Steel Soc. (1985)
5. Eser, S. and Jenkins, R. G., *Carbon,* 27(6):877-887 (1989)
6. Akezuma, M. et al, *Carbon,* 25(4):517-522 (1987)
7. Sato, Y, Kitano, T., Inagaki, M., and Sakai, M., *Carbon,* 28(1):143-148 (1990)
8. Charette, A. et al, *Carbon,* 29(7):1015-1024 (1991)
9. Kochling, K. H., McEnaney, B., Muller, S. and Fitzer, E., *Carbon,* 24(2):246-247 (1986)
10. *High-Purity Graphite Products for the Semiconductor Industry,* Bulletin from The Carbon/Graphite Group Inc., St. Marys, PA 15857
11. *A Machinist's Guide for Graphite,* Bulletin from the Graphite Materials Div., Carbone of America, St. Marys, PA 15857
12. *Graphite, Refractory Material,* Bulletin from Le Carbone-Lorraine, Gennevilliers 92231, France
13. *Carbon/Graphite Properties,* Bulletin from the Graphite Materials Div., Carbone of America, St. Marys, PA 15857
14. *Carbon Products,* Bulletin from The Carbon/Graphite Group Inc., St. Marys, PA 15857
15. Nystrom, W. A., *High-Quality, Cost-Effective Graphites for the Metal Casting Industry,* Bulletin from the Graphite Materials Div., Carbone of America, St. Marys, PA 15857 (1988)
16. *Products for the Semiconductor Industry,* Bulletin from Ringsdorff, D-5300 Bonn-2, Germany (1988)
17. Mantell, C. L., *Carbon and Graphite Handbook,* 391-424, Interscience Publishers, New York (1968)
18. Engle, G. B. and Eatherly, W. P., *High Temperatures- High Pressures,* 4:119-158 (1972)
19. Gray. W. J. and Pitner, A. L., *Carbon,* 9:699-710 (1971)

6

Vitreous Carbon

1.0 GENERAL CONSIDERATIONS

The molded graphites, reviewed in the previous chapter, are derived from precursors that graphitize readily, such as petroleum cokes and coal-tar pitch. They exhibit varying degrees of anisotropy and have characteristics and properties that, in some cases, can be very similar to those of the ideal graphite crystal.

The materials reviewed in this chapter form another distinctive group of carbon materials: the vitreous carbons. Like molded graphite, vitreous (glassy)carbon is processed by the carbonization (pyrolysis) of an organic precursor. Unlike most molded graphites, it does not graphitize readily and has characteristics and properties that are essentially isotropic. The difference between these two classes of materials stems from different precursor materials.

Vitreous carbon has a structure that is more closely related to that of a glassy material (i.e., non-crystalline), with high luster and glass-like fracture characteristics, hence the name vitreous (or glassy). Vitreous carbon is also frequently called polymeric carbon since it derives mostly from the carbonization of polymeric precursors. In the context of this book, the terms vitreous, glassy and polymeric carbons are synonymous.

Vitreous carbon is a relatively new material which was developed in the 1960's. It has some remarkable properties, such as high strength, high resistance to chemical attack, and extremely low helium permeability.

122

Vitreous carbon has carved its own niche in the carbon industry (although on a much smaller scale than the molded graphites) with some important industrial applications and a growing market.

2.0 PRECURSORS AND PROCESSING

Vitreous carbon is obtained by the carbonization of organic polymers, commonly known as plastics. The types of polymers and the carbonization process and mechanism are reviewed in broad terms in Ch. 3, Sec. 2.

2.1 Polymeric Precursors

To be a suitable precursor for vitreous carbon, a polymer must have the following characteristics:[1]

- The structure of the molecule must be three-dimensionally cross-linked.

- Carbonization must take place in the solid state, without mesophase formation, and result in the formation of a char (as opposed to coke).

- The molecular weight and the degree of aromaticity (i.e., the number of benzene rings) must be high to provide a relatively high carbon yield.

The following polymeric precursors are, or have been, used in production or experimentally.[2][3]

Polyfurfuryl alcohol: Polyfurfuryl alcohol is a thermosetting resin which is obtained by the polymerization of the furfuryl-alcohol monomer catalyzed with maleic acid. Its chemical structure is the following:

$$
\left[
\begin{array}{c}
\mathrm{HOH_2C} \qquad\ \ \mathrm{O} \\
\diagdown \ \diagup \diagdown \\
\mathrm{C} \qquad \mathrm{C\text{-}H} \\
\parallel \qquad\ \parallel \\
\mathrm{H\text{-}C} - \mathrm{C\text{-}H}
\end{array}
\right]_n
$$

In addition to being a precursor for vitreous carbon, polyfurfuryl alcohol is also a common impregnant for graphite electrodes (see Ch. 5).

Phenolics. Phenolics, typically $(C_{15}O_2H_{20})_n$, are a class of polymers which are obtained by a *condensation reaction,* usually between phenol and formaldehyde, with elimination of water. The monomer has the following chemical structure:

When heated to about $250°C$, extensive cross-linking occurs and a hard, rigid, and insoluble polymer is formed.

Polyimide. The polymer polyimide, $(C_{22}H_{10}O_5N_2)_n$, in the form of a film (known then as Kapton or Novax), is unusual in the sense that it carbonizes to form a char which becomes a well-oriented graphite after heat-treatment to $3000°C$ (see Ch. 4, Sec. 3.5).[4][5]

Polyacrylonitrile. The polymer polyacrylonitrile (PAN) is used mostly for the production of carbon fibers.

Cellulose. Cellulose is the major constituent of wood and other plants. Cotton and rayon are almost pure cellulose and the latter is a precursor for carbon fibers. The pyrolysis of cellulose, like that of PAN, is reviewed in Ch. 8.

Others. Other polymers are being investigated as potential precursors for glassy carbon, such as polyvinylidene chloride $(CH_2CCl_2)_n$, polyvinyl alcohol (CH_2CHOH), polyphenylene oxide and aromatic epoxy. The latter two compounds have a high carbon yield.

2.2   Processing and Carbonization

Molding. The precursor polymer is often combined with other materials such as solvents to obtain the desired molding and carbonization characteristics. Some of these compositions and processes are considered proprietary by the manufacturers.

The precursor compound is heated and molded to the desired shape by standard plastic processing, i.e., injection molding, extrusion, or compression molding. Since carbonization is accompanied by a large shrinkage, the dimensions of the molded part must be larger and calculated so the part after carbonization will be close to net-shape and machining will be minimized.

Carbonization. The molded part is carbonized by slow heating in a reducing or inert environment. A typical heating cycle is: 1 - 5°C/min to 800°C and 5 - 10°C/min from 800 - 1000°C. The heating rate is a function of the rate of diffusion of the volatile by-products of pyrolysis. This diffusion rate must be slow to avoid disruption and rupture of the carbon structure: the thicker the part, the slower the rate. For very thin parts, i.e., 10 μm or less, only a few minutes may be needed; for thicker parts, carbonization may take weeks. For that reason, wall thickness is usually limited to less than 4 mm.

The exact carbonization mechanism is still conjectural but appears to be related to the degree of cross-linking of the precursor polymer. A proposed mechanism for the carbonization of polyfurfuryl alcohol is shown in Fig. 6.1.[6] The volatile compounds, H_2, CO, CO_2, CH_2, and H_2O, slowly diffuse out and, when 1300°C is reached, the material is essentially all carbon.

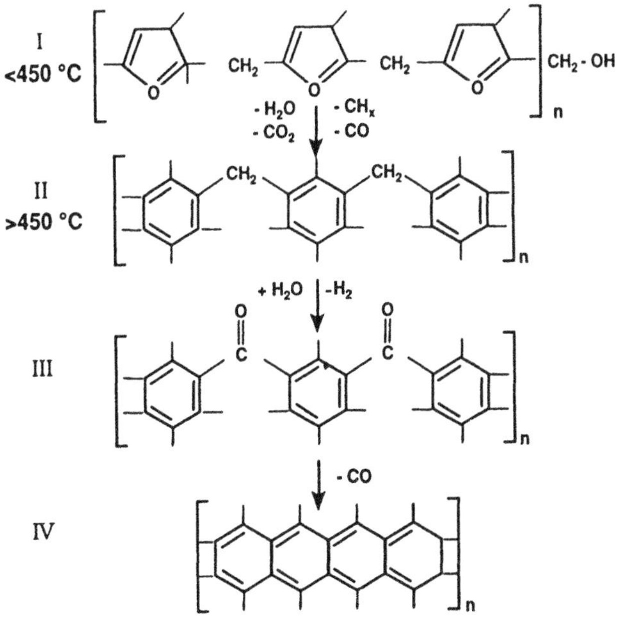

Figure 6.1. Proposed mechanism of carbonization of furfuryl alcohol.[6]

During this carbonization cycle, the cross-linked polymeric chains do not break down and, unlike coke materials, do not go through a mesophase. The carbonization of phenolic polymers follows a similar path.[6]

Weight Loss and Dimensional Changes. The weight loss resulting from the elimination of non-carbon constituents tapers off after the temperature reaches 1000°C, as shown in Fig. 6.2. The phenol formaldehyde precursor has a total weight loss approaching 40% with a carbon yield of 85%.[6]

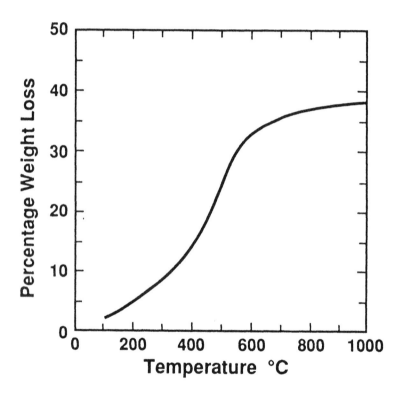

Figure 6.2. Weight loss of phenolic resin during carbonization.[6]

The dimensional changes are shown in Fig. 6.3. Volumetric shrinkage is approximately 50%. This percentage is higher than the percent weight loss and can be accounted for by the higher density of the vitreous carbon (~ 1.47 g/cm³) compared to that of the phenolic resin (~ 1.2 g/cm³).

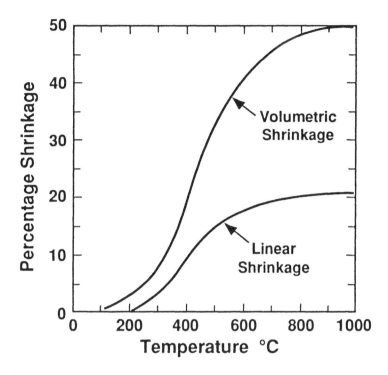

Figure 6.3. Dimensional changes of phenolic resin during carbonization.[6]

2.3 Graphitization

Vitreous carbon is essentially a char and generally does not graphitize to any extent. Heat-treating at 1800°C produces a material with a interlayer spacing (d) of 0.36 nm and a crystallite size (L_c) of 1.5 nm. The crystallite size is estimated from the broadening of the (110) and (002) lines of the x-ray diffraction patterns.

These patterns are not appreciably altered after heating the vitreous carbon to 2750°C, in contrast with the pattern of heat-treated petroleum coke which indicates a well-graphitized material as shown in Fig. 6.4.[7] The crystallite size (L_c) of the vitreous carbon remains small (up to 3 nm) and the interlayer spacing (d) shows only a slight reduction to a minimum of 0.349 nm.[2]

Heat treatment up to 3000°C causes a volumetric expansion of approximately 5%.[1]

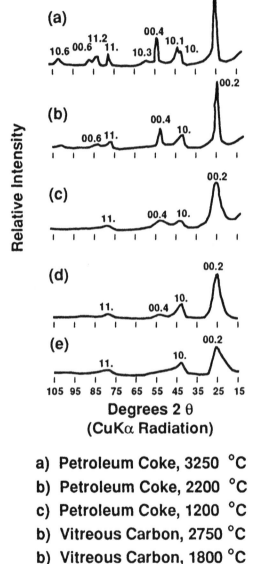

a) Petroleum Coke, 3250 °C
b) Petroleum Coke, 2200 °C
c) Petroleum Coke, 1200 °C
b) Vitreous Carbon, 2750 °C
b) Vitreous Carbon, 1800 °C

Figure 6.4. X-ray diffraction patterns of petroleum coke and vitreous carbon.[7]

3.0 STRUCTURE AND PROPERTIES OF VITREOUS CARBON

3.1 Structure

After carbonization, the residual material is essentially all carbon and, from the structural standpoint, is a vitreous carbon. A substance is considered vitreous (or glassy) when it has no crystalline long-range order, i.e., when the arrangement of its molecular constituents is only a few times the size of each constituent.[8] In the case of vitreous carbon, this means small, randomly oriented crystallites (L_c up to ~ 3.0 nm). Within each crystallite, the interatomic distances deviate from those of the ideal graphite crystal by more than 5% in both the basal plane (ab directions) and between planes (c direction).

This random structure, characteristic of vitreous carbon, is believed to have the form of an extensive and stable network of graphitic ribbons, shown in the simplified schematic of Fig. 6.5.[2] These ribbon-like tangled aromatic molecules are cross-linked by carbon-carbon covalent bonds with varying bond energies. Some of these bonds may be highly strained.[9]

Figure 6.5. Proposed model of the ribbon structure of glassy carbon.[2]

It has been suggested that the structure of vitreous carbon includes, in addition to the graphitic sp^2 (trigonal) bonds, some sp^3 (tetragonal) bonds which are characteristic of diamond bonding. This would possibly contribute to the isotropy, the high strength, and the hardness of the material (see Ch. 2, Sec. 3.0).[2][7] The presence of the diamond structure in a similar material, diamond-like carbon, is well established (see Ch. 14).

These two factors, ribbon network and sp^3 bonds, would prevent further ordering of the structure, regardless of the graphitization temperature.

3.2 Porosity

Vitreous carbon has low density (approximately two-thirds that of the ideal graphite crystal) which implies that it has high porosity. Yet the permeability of the material is exceptionally low. Helium permeability for instance, measured by the vacuum-drop method, is only 10^{-11} cm²/s. This means that the pores are extremely small and essentially inaccessible to helium. Pore diameters of 0.1 to 0.3 nm are reported with an average of 0.25 nm after heat-treatment to 220°C and 0.35 nm to 3000°C.[1][2]

This extremely fine-pore structure gives vitreous carbon the characteristics of a molecular sieve and allows the absorption of some very small molecules as shown in Table 6.1.[1] This table shows a minimal absorption for water (in spite of the small diameter of its molecule). This contradicts the results of Fitzer[6] who found the absorption to be similar to that of methanol.

Table 6.1. Adsorption of Selected Molecules by Vitreous Carbon

Compound	Gas-kinetic molecular diam., nm	% increase by wt. after 20 days
Water	0.289	0.03
Methanol	0.376	1.34
Isopropanol	0.427	1.16
Isobutanol	0.475	0

3.3   Types of Vitreous Carbon

Vitreous carbon can be produced in three basic types which have essentially the same microstructure, but different macrostructures: solid (or monolithic), foam (or reticulated), and spheres (or particles). Each type is reviewed in the next three sections.

4.0   SOLID VITREOUS CARBON

4.1   Physical, Mechanical, and Thermal Properties

Because of its random structure, vitreous carbon has properties that are essentially isotropic. It has low density and a uniform structure which is generally free of defects. Its hardness, specific strength, and modulus are high. Its properties (as carbonized and after heat-treatment to 3000°C) are summarized in Table 6.2.[1] The table includes the properties of a typical molded graphite and of pyrolytic graphite for comparison (see Chs. 5 and 7). The mechanical properties of vitreous carbon are generally higher and the thermal conductivity lower than those of other forms of carbon.

The thermal conductivity, coefficient of thermal expansion, and electrical resistivity vary with temperature as shown in Fig. 6.6.[1]

Table 6.2.   Physical and Mechanical Properties of Vitreous Carbon and Other Carbon Materials at 25°C

Properties	Vitreous (As Is)	Carbon (Graph.)	Molded Graphite	Pyrolytic Graphite
Density, g/cm^3	1.54	1.42	1.72-1.90	2.10-2.24
Flexural strength, MPa	210	260	10-100	80 - 170 (c)
Compressive strength, MPa	580	480	65 - 89	
Young's modulus of elasticity, GPa	35	35	5 - 10	28 - 31
Vickers hardness, HV$_1$	340	230	40 - 100	240 - 370
Coef. of thermal expansion 20 - 200°C, m/m·K x 10^{-6}	3.5	2.6	3.2-5.7	0 (ab) 15 - 25 (c)
Thermal conductivity, W/m·K	4.6	6.3	31 - 159	1 - 3 (c) 190 - 390 (ab)

Note: (ab) = ab directions; (c) = c direction

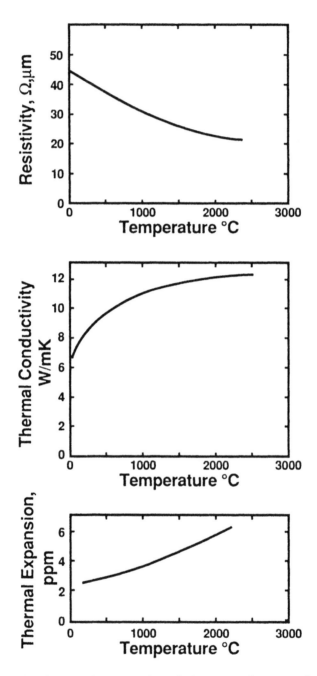

Figure 6.6. Variations in the properties of vitreous carbon as a function of temperature.[1]

4.2   Chemical Properties

In most cases, the chemical properties of vitreous carbon are similar to those of the graphite crystal, reviewed in Ch. 3, Sec. 7. Since the material has low permeability, is essentially non-porous and free of surface defects, and can be made with very low impurities, its resistance to chemical attack is generally excellent and is one of its outstanding characteristics. In many instances, it is far more chemically resistant than other forms of carbon, such as molded or pyrolytic graphites.

The presence of impurities (ashes) is very critical. For instance, the rate of oxidation increases by approximately an order of magnitude when the amount of impurities increases by the same ratio, as shown in Fig. 6.7.[7] This figure also shows that the oxidation of vitreous carbon is much less than that of molded graphite.

Vitreous carbon does not react with nitric, sulfuric, hydrofluoric, and chromic acids. It is not attacked by halogens such as bromine, even at high temperatures, as opposed to other graphitic materials which are attacked rapidly.[8] Its rate of reaction with various reagents is shown in Table 6.3.

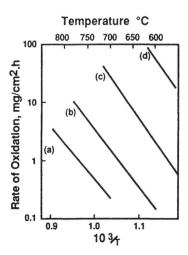

a) Vitreous carbon, 20 ppm impurities
b) Vitreous carbon, 200 ppm impurities
c) Fine-grain graphite
d) Baked carbon

Figure 6.7. Oxidation rate of vitreous carbon and other carbon products.[7]

Table 6.3. Reaction Rate of Vitreous Carbon with Various Reagents[7]

Reagent	Temperature °C	Rate of weight loss $(g/cm^2 \cdot hr)$*
Steam	570 - 580	0.05
Carbon dioxide	1000	$0.1\text{-}0.5 \times 10^{-3}$
Carbon dioxide	2500	0.8
10% Oxygen, 90% Argon	2500	1.8
10% Hydrogen, 90% Argon	2500	0.07
Fused potassium bisulfate	500	2.4×10^{-4}
Fused boric oxide	580	$<5 \times 10^{-6}$
Fused caustic soda	320	$<5 \times 10^{-6}$

* Apparent surface

4.3   Shrinkage and Machining

As mentioned in Sec. 2.2 above, the precursor polymer shrinks considerably and the design of the molded part must accommodate this shrinkage to obtain a final product close to the desired shape. This is sometimes difficult to achieve and a machining operation is often necessary. Because of the high hardness of vitreous carbon, it should be machined with diamond tools.

4.4   Applications

Chemical Applications. Many applications of vitreous carbon are based on its outstanding chemical resistance. These applications include vessels for chemical processing and analytical chemistry such as crucibles, beakers, boats, dishes, reaction tubes, lining for pressure vessels, etc.[10]

Metallurgical and Glass Applications. Vitreous carbon reacts to molten metals that form carbides readily such as the elements of groups IV, V and VI, i.e., Ti, Zr, Hf, Nb, Ta, W. However, it is inert to and not wetted by other molten metals and, for that reason, it is used widely as crucibles for the melting of noble metals and special alloys, particularly in dental technology. The excellent thermal shock resistance permits very high rates of heating and cooling. Vitreous carbon is not wetted by glass and is used as mold for lenses and other glass products.

Battery Electrodes. The chemical inertness and good electrical conductivity of vitreous carbon makes it a potentially excellent material for acid-battery electrodes.[11][12]

5.0   VITREOUS CARBON FOAM

Vitreous carbon can be produced in the form of an open-pore foam, known as reticulated vitreous carbon (RVC). The precursor materials are essentially the same polymers used for solid vitreous carbon, except that they are foamed prior to carbonization. The carbonization process is similar. The open-pore structure of the material is shown in Fig.6.8.[13]

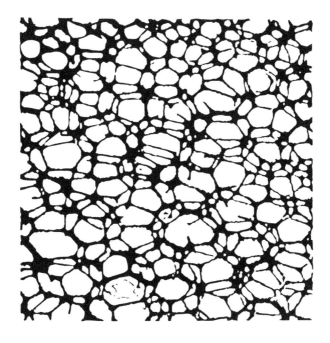

Figure 6.8. Open-pore structure of vitreous carbon foam.[13]

5.1   Characteristics and Properties

Vitreous carbon foam is produced in several pore sizes, usually described as number of pores per inch (ppi). Commercially available foams are respectively 60, 100, and 200 ppi (24, 39 and 78 pores per cm). These foams have low density, with relatively even pore distribution. Their properties are listed in Table 6.4.[14]

Vitreous-carbon foam is very susceptible to oxidation due its large surface area. Any application involving an oxidizing atmosphere above 500°C should not be considered.

Table 6.4. Properties of Vitreous Carbon Foam

Bulk void volume, %	97
Bulk density, g/cm^3	0.05
Strut density, g/cm^3	1.49
Strut resistivity, 10^{-4} ohm·cm	50
Crushing strength, MPa (function of pore size)	0.07 - 3.4
Surface area, m^2/g	1.62

5.2   Applications

Electrodes. Its chemical inertness, its wide range of usable potential (1.2 to -1.0 V vs. SCE) and the hydrodynamic and structural advantages of its open-pore foam structure make vitreous carbon foam an attractive material for electrodes for lithium-ion and other types of batteries, with many potential applications in electrochemistry.[13][15][16]

High-Temperature Thermal Insulation. A potential application of vitreous-carbon foam is high-temperature thermal insulation in vacuum or non-oxidizing atmosphere. Several factors combine to make this structure an excellent thermal insulator: (a) the low volume fraction of the solid phase, which limits conduction; (b) the small cell size, which virtually eliminates convection and reduces radiation through repeated absorption/reflection at the cell walls; and (c) the poor conductivity of the enclosed gas (or vacuum). An additional advantage is its excellent thermal-shock resistance due to its relatively low modulus compared to the bulk material. Very high thermal gradients can be tolerated.

Adsorption of Hydrocarbons and Other Gases. In the activated form, vitreous-carbon foam could replace activated-carbon granules without the requirement of a container (see Ch. 10, Sec. 4.0). Potential uses are in emission control and recovery.

Other Applications. Vitreous-carbon foam is being considered as a filter for diesel particulates and for the filtration of non-carbide-forming molten metals.

6.0 VITREOUS CARBON SPHERES AND PELLETS

Vitreous carbon in the form of microspheres or pellets has a number of applications, especially in the field of catalytic supports.

6.1 Processing

A typical process for the production of vitreous-carbon spheres is represented schematically in Fig. 6.9.[17] The precursor is a partially polymerized polymer such as furfuryl alcohol, catalyzed with p-toluene sulfonic acid and mixed with acetone to obtain the proper viscosity for atomization.[18] A pore former is added which can be an organic material with a high boiling point or sub-micron solid particles such as carbon black. Atomization occurs in the thermal reactor shown schematically in Fig. 6.10.[17] The curing time is very brief because of the small size of the particles ($\sim$ 45 μm).

The microspheres are then heat-treated from 530 to 1330°C. If required, they can be partially oxidized to create micro- and transitional-pores.

6.2 Applications

Catalytic Support. Vitreous carbon spheres are being considered as catalyst supports for iron and other metals. The material may offer some important advantages over other forms of carbon, such as lower inorganic impurities (which may poison the catalyst) and a more uniform pore structure. The activation mechanism and the properties and characteristics of catalytic materials are reviewed in greater detail in Ch. 10, Sec. 4.0.

Other Applications: Other applications include foams, low density fillers for plastics and high-temperature thermal insulation.

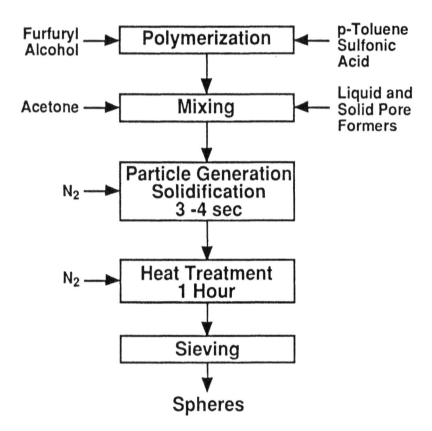

Figure 6.9. Processing flow-chart for vitreous carbon spheres.[17]

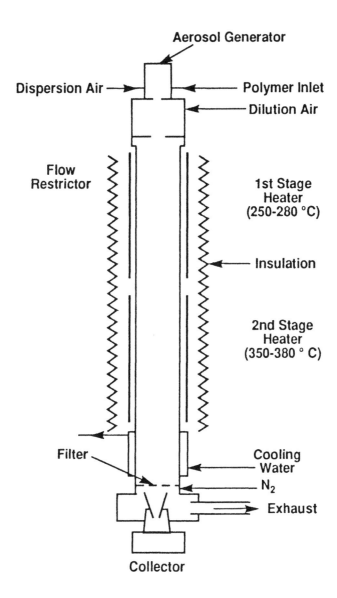

Figure 6.10. Schematic of vitreous-carbon production apparatus.[17]

REFERENCES

1. Dubgen, R., *Glassy Carbon - A Material for Use in Analytical Chemistry*, Publication of Sigri, D-8901 Meitingen, Germany (1985)
2. Jenkins, G. M. and Kawamura, K., *Polymeric Carbons*, Cambridge Univ. Press, Cambridge, UK (1976)
3. Fitzer, E., Schaefer, W., and Yamada, S., *Carbon*, 7:643-648 (1969)
4. Inagaki, M., et al, *Carbon*, 27(2):253-257 (1989)
5. Inakagi, M. et al, *Carbon*. 29(8):1239-1243 (1991)
6. Fitzer, E. and Schaefer, W., *Carbon*, 8:353-364 (1970)
7. Cowlard, F. and Lewis J., *J. of Mat. Sci.* 2:507-512 (1967)
8. Doremus, R., *Glass Science*, John Wiley & Sons, New York (1973)
9. Jenkins, G. M. and Kawamura, K., *Nature*, 231:175-176 (May 21, 1971)
10. Lewis, J. C., Redfern, B., and Cowlard, F. C., *Solid-State Electronics*, 6:251-254, Pergammon Press (1963)
11. Lausevic, Z. and Jenkins, G. M., *Carbon*, 24(5):651-652 (1986)
12. Van der Linden, W. E., and Dieker, J. W., *Analytica Chimica Acta*, 119:1-24 (1980)
13. Wang, J., *Electrochimica Acta*, 26(12):1721-1726 (1981)
14. *Reticulated Vitreous Carbon*, Brochure form ERG, Oakland, CA 94608 (1976)
15. Sherman, A. J., Tuffias, R. H., and Kaplan, R. B., *Ceramic Bull.*, 70(6):1025-1029 (1991)
16. Sherman, A. J. and Pierson, H. O., *Ultrastructures for Cold Cathode Emitters,* Final Report (ULT/TR-89-6762), Air Force Electronic Systems Division, Hanscom AFB, MA (April 1989).
17. Levendis, Y. and Flagan, R., *Carbon*, 27(2):265-283 (1989)
18. Moreno-Castilla, C., et al, *Carbon*, 18:271-276 (1980)

7

Pyrolytic Graphite

1.0 GENERAL CONSIDERATIONS

The production of molded graphite and vitreous carbon, described in the previous two chapters, relies on the carbonization (pyrolysis) of a solid, inorganic substance such as coal-tar pitch, petroleum fractions or polymers. This chapter is a review of another type of carbon material, produced by a fundamentally different process that is based on a gaseous precursor instead of a solid or liquid. The process is known as chemical vapor deposition (CVD) and the product as pyrolytic carbon or graphite, sometimes referred to as pyrocarbon or pyrographite. To simplify, in this chapter the material will be referred to as pyrolytic graphite, regardless of the degree of graphitization.

Pyrolytic graphite is different from another standpoint: although produced in bulk form, its main use is in the form of coatings, deposited on substrates such as molded graphite, carbon fibers, or porous carbon-carbon structures. As such, it is part of a composite structure and is not as readily identifiable as other forms of carbon. It is similar in this respect to CVD diamond and diamond-like carbon (DLC) described in Chs. 13 and 14.

Pyrolytic graphite is the only graphitic material that can be produced effectively as a coating. The coating can be made sufficiently thick that, after removing the substrate, a free-standing object remains.

Pyrolytic graphite is a key element in the technology of carbon and is used extensively in the coating of specialty molded graphites and in the processing of carbon-carbon components.

1.1   Historical Perspective

The CVD of carbon materials is not new. As mentioned in the pioneer work of Powell, Oxley, and Blocher,[1] its first practical use was developed in the 1880's in the production of incandescent lamps to improve the strength of filaments by carbon deposition and a patent was issued over a hundred years ago, covering the basis of the CVD of carbon.[2]

The CVD process developed slowly in the next fifty years, and was limited mostly to pyro and extraction metallurgy, and little work was done on graphite deposition.

It is only since the end of World War II that the CVD of graphite began to expand rapidly as researchers realized the potential of this technique for the formation of coatings and free-standing shapes. The importance and impact of pyrolytic graphite have been growing ever since.

1.2   The Chemical Vapor Deposition Process

CVD is now a well-established process that has reached major production status in areas such as semiconductors and cutting tools. It is a vapor-phase process which relies on the chemical reaction of a vapor near or on a heated surface to form a solid deposit and gaseous by-products. The process is very suitable to the deposition of carbon, as reviewed below.[3]

1.3   Pyrolytic Graphite as a Coating

Although, as mentioned above, pyrolytic graphite is used by itself as free-standing structures such as crucibles or rocket nozzles (see Sec. 4.0), its major use is in the form of coatings on substrates such as molded graphite, carbon foam, carbon fibers, metals, and ceramics.

Composite Nature of Coatings. The surfaces of many materials exposed to the environment are prone to the effects of abrasion, corrosion, radiation, electrical or magnetic fields, and other conditions. These surfaces must have the ability to withstand these environmental conditions and/or provide certain desirable properties such as reflectivity, semi-conductivity, high thermal conductivity, or erosion resistance.

To obtain these desirable surface properties, a coating is deposited on the bulk material to form a composite in which bulk and surface properties may be very different.[4]

Table 7.1 summarizes the surface properties that may be obtained or modified by the use of pyrolytic graphite coatings.

Table 7.1. Material Properties Affected by Pyrolytic Graphite Coatings

Electrical	Resistivity
Optical	Reflectivity
Mechanical	Wear
	Friction
	Hardness
	Adhesion
	Toughness
Porosity	Surface area
	Pore size
	Pore volume
Chemical	Diffusion
	Corrosion
	Oxidation

2.0 THE CVD OF PYROLYTIC GRAPHITE

The CVD of pyrolytic graphite is theoretically simple and is based on the thermal decomposition (pyrolysis) of a hydrocarbon gas. The actual mechanism of decomposition however is complex and not completely understood.[5] This may be due in part to the fact that most of the studies on the subject of hydrocarbon decomposition are focused on the improvement of fuel efficiency and the prevention of carbon formation (e.g., soot), rather than the deposition of a coating.

Although many studies of the CVD of graphite have been carried out, a better understanding of the pyrolysis reactions, a more accurate prediction of the results, and more complete experimental, thermodynamic, and kinetic investigations are still needed.

2.1   Thermodynamics and Kinetics Analyses

The CVD of pyrolytic graphite can be optimized by experimentation. The carbon source (hydrocarbon gas), the method of activating the decomposition reaction (thermal, plasma, laser, etc.), and the deposition variables (temperature, pressure, gas flow, etc.) can be changed until a satisfactory deposit is achieved.   However, this empirical approach may be too cumbersome and, for more accurate results, it should be combined with a theoretical analysis.

Such an analysis is a valuable step which, if properly carried out, predicts what will happen to the reaction, what the resulting composition of the deposit will be (i.e., stoichiometry), what type of carbon structure to expect, and what the reaction mechanism (i.e., the path of the reaction as it forms the deposit) is likely to be.   The analysis generally includes two steps:

1. The calculation of the change in the free energy of formation for a given temperature range; this is a preliminary, relatively simple step which provides information on the feasibility of the reaction.

2. The minimization of the free energy of formation which is a more complete analysis carried out with a computer program.

2.2   ΔG Calculations and Reaction Feasibility

Thermodynamics of CVD Carbon. The CVD of carbon (as all CVD reactions) is governed by two factors: *(a) thermodynamics*, that is the driving force which indicates the direction the reaction is going to proceed (if at all), and *(b) kinetics*, which defines the transport process and determines the rate-control mechanism, i.e., how fast it is going.

Chemical thermodynamics is concerned with the interrelation of various forms of energy and the transfer of energy from one chemical system to another in accordance with the first and second laws of thermodynamics.   In the case of CVD, this transfer occurs when the gaseous compounds, introduced in the deposition chamber, react to form the carbon deposit (and by-products gases).

ΔG Calculation: The first step is to ensure that the desired CVD reaction will take place in a given temperature range. This will happen if the thermodynamics is favorable, that is, if the transfer of energy (i.e., the free-energy change of the reaction, known as ΔG_r) is negative. To calculate ΔG_r, it is necessary to know the thermodynamic properties of each component, specifically their free-energy of formation (also known as Gibbs free energy), ΔG_f. The values of ΔG_r of the reactants and products for each temperature can be obtained from thermodynamic-data tables such as the JANAF Thermochemical Tables and others.[6][7]

It should be noted that the negative free-energy change is a valid criterion for the feasibility of a reaction only if the reaction as written contains the major species that exist at equilibrium.

2.3 Minimization of Gibbs Free Energy

Experimentation shows that the best, fully dense, and homogeneous carbon deposits are produced at an optimum negative value of ΔG. For smaller negative values, the reaction rate is very low and, for higher negative values, vapor-phase precipitation and the formation of soot can occur. Such factors are not revealed in the simple free-energy change calculation. A more complete analysis is often necessary.

A method of analysis is the minimization of the Gibbs free energy, a calculation based on the rule of thermodynamics which states that a system will be in equilibrium when the Gibbs free energy is at a minimum. The objective then is the minimization of the total free energy of the system and the calculation of equilibria at constant temperature and volume or constant pressure. It is a complicated and lengthy operation but, fortunately, computer programs are now available that simplify the task considerably.[8][9]

These programs provide the following information:

- The composition and amount of deposited material that is theoretically possible at a given temperature, pressure, and concentration of input gases

- The existence of gaseous species and their equilibrium partial pressures

- The possibility of multiple reactions with the inclusion of the substrate as a possible reactant

All of this is valuable information which can be of great help. Yet, it must be treated with caution since, in spite of all the progress in thermodynamic analysis, the complexity of many reactions in the CVD of carbon, and the fact that these calculations are based on chemical equilibrium which is rarely attained in CVD reactions, make predictions relying on thermodynamic calculations alone still questionable.

It follows that, in order to provide a reliable and balanced investigation, it is preferable to combine the theoretical calculations with an experimental program. Fortunately, carbon deposition experiments are relatively easy to design and carry out without the need for expensive equipment, and results can usually be obtained quickly and reliably.

2.4   CVD Reactions for the Deposition of Pyrolytic Graphite

The CVD reactions to deposit pyrolytic graphite are based on the thermal decomposition (pyrolysis) of hydrocarbons. The most common precursor is methane (CH_4), which is generally pyrolyzed at 1100°C or above, over a wide range of pressure from about 100 Pa (0.001 atm) to 10^5 Pa (1 atm). The reaction in a simplified form is as follows:[1][10][11]

Eq. (1) $CH_4 \rightarrow C + 2H_2$

Other common precursors are ethylene (C_2H_6) and acetylene (C_2H_2).[1][12] Acetylene can also be decomposed at lower temperature (300 - 750°C) and at pressures up to 1 atm, in the presence of a nickel catalyst.[12] Another common precursor is propylene (C_3H_6) which decomposes in the 1000 - 1400°C temperature range at low pressure (~ 1.3 x 10^4 Pa or 100 Torr).[13]

The activation energies for the decomposition of these precursor gases still are not accurately known and show a considerable scatter. The reported values are as follows:

Methane	78 - 106 kcal/g·mole
Ethane	60 - 86 kcal/g·mole
Acetylene	30 - 50 kcal/g·mole

Deposition Mechanism: The pyrolysis of a hydrocarbon, such as shown in reaction equation (1), is actually a series of more complex reactions involving molecules of gradually increasing size. A possible mechanism of deposition of pyrolytic graphite is deduced from a series of experiments carried out in the apparatus shown in Fig. 7.1.[5]

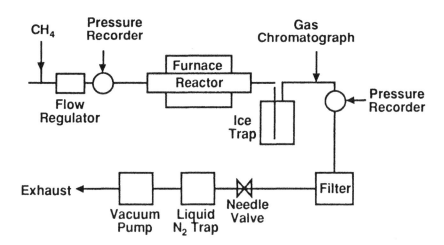

Figure 7.1. Schematic of experimental apparatus for the production of pyrolytic graphite.[5]

In this study, a spectrographic analysis of the by-products of the decomposition of methane revealed the presence of large amounts of acetylene, ethylene, and benzene, plus a variety of compounds consisting mostly of the polyaromatic hydrocarbons (PAH) such as naphthalene, anthracene, phenantrene, acenaphthylene, pyrene, and fluoranthene, in addition to the deposited pyrolytic graphite. Some of these compounds form the soot and tar-like deposits which are often observed on the wall of CVD reactors during carbon deposition.

It is generally agreed that the following simplified deposition sequence is taking place:[5][14]

Methane → Benzene → Polyaromatic hydrocarbons → Carbon

2.5 Deposition Systems and Apparatus

A common CVD apparatus for the deposition of pyrolytic graphite is the so-called cold-wall reactor. This reactor does not require a furnace since the substrate to be coated is heated directly, usually by induction heating.

The decomposition reactions for the deposition of pyrolytic graphite are endothermic, i.e., they absorb heat. As a result, deposition takes place preferentially on the surfaces where the temperature is the highest, in this case the substrate, while the cooler walls of the reactor remain essentially uncoated.

A simple laboratory type reactor for pyrolytic-graphite deposition is shown in Fig. 7.2.[3] The substrate is a molded-graphite disk which is rotated to improve deposition uniformity. It is heated by a high-frequency (450 kHz) induction coil and deposition occurs at low pressure (500 Pa). Temperature is monitored and controlled by a sheathed thermocouple and corroborated by an optical pyrometer.

Production systems of a similar basic design now reach considerable size, with CVD furnaces 1.2 m in diameter and over 2 m high commercially available.

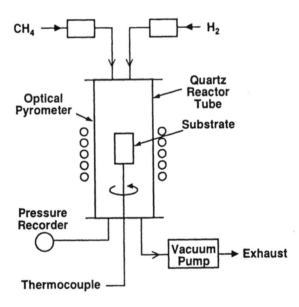

Figure 7.2. Schematic of a cold-wall reactor for the production of pyrolytic graphite.[3]

2.6   Chemical Vapor Infiltration (CVI)

Chemical vapor infiltration (CVI) is a special CVD process in which the gaseous reactant infiltrates a porous material such as an inorganic open foam or a fibrous mat or weave. The deposition occurs on the fiber (or the foam), and the structure is gradually densified to form a composite.[15]

CVI has the same chemistry and thermodynamics as conventional CVD, but the kinetics is different since the reactants have to diffuse inward through the porous structure and the by-products of the reaction have to diffuse out.[16] The process is used extensively in the production of carbon-carbon materials, reviewed in Ch. 9.[17]

2.7   Fluidized-Bed CVD

Fluidized-bed CVD is a special technique which is used primarily in coating particles such as nuclear fuel. A flowing gas imparts quasi-fluid properties to the particles. Fig. 7.3 shows a typical fluidized-bed CVD reactor.[3]

The fluidizing gas is usually methane, helium, or another non-reactive gas. Factors to consider to obtain proper fluidization are the density and size of the particles to be coated, and the velocity, density, and viscosity of the gases.[18] If the velocity is too low, the particles will fall into the gas inlet; if it is too high, they will be blown out of the bed. Heavy or large objects may require suspension in the bed.

The gas velocity, V_m, is given by the following relationship:

$$V_m = d^2 (\rho_p - \rho_g) G / 1650\mu \qquad \text{[for } dV_o \rho_g / \mu < 20]$$

where: d = particle diameter
            ρ_p = particle density
            ρ_g = gas density
            G = acceleration of gravity
            V_o = superficial gas velocity
            μ = gas viscosity

The major applications of pyrolytic carbon deposited by fluidized bed are found in the production of biomedical components such as heart valves and in the coating of uranium carbide and thorium carbide nuclear-fuel particles for high temperature gas-cooled reactors, for the purpose of containing the products of nuclear fission. The carbon is obtained from the decomposition of propane (C_3H_8) or propylene (C_3H_6) at 1350°C, or of methane (CH_4) at 1800°C.[1] Its structure is usually isotropic (see Sec. 3.5).

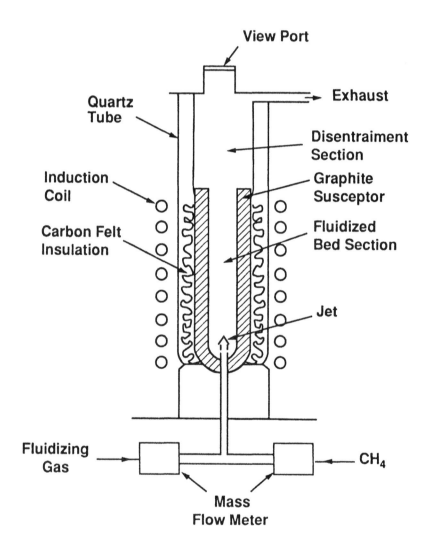

Figure 7.3. Schematic of a fluidized-bed CVD reactor for the deposition of pyrolytic graphite.[3]

2.8   Plasma CVD

The deposition of graphite can also be obtained by plasma CVD, with the following characteristics:[19]

- Gases: propylene-argon or methane-argon
- Plasma: radio frequency (RF) at 0.5 MHz
- Pressure: <1300 Pa
- Temperature: 300 - 500°C

In a plasma-activated reaction, the substrate temperature can be considerably lower than in thermal CVD. This allows the coating of thermally sensitive materials. The characteristics, and properties of the coating are similar to those of coatings deposited at higher temperatures (>1000°C).

Plasma activation is also used extensively in the deposition of polycrystalline diamond and diamond-like carbon (DLC). It is reviewed in more detail in Chs. 13 and 14.

3.0   STRUCTURE OF PYROLYTIC GRAPHITE

3.1   The Various Structures of Pyrolytic Graphite

Pyrolytic graphite is an aggregate of graphite crystallites which have dimensions (L_c) that may reach several hundred nm (see Ch. 3, Sec. 2). It has a turbostratic structure, usually with many warped basal planes, lattice defects, and crystallite imperfections.

Within the aggregate, the crystallites have various degrees of orientation. When they are essentially parallel to each other, the nature and the properties of the deposit closely match that of the ideal graphite crystal.

The structure of a pyrolytic graphite deposit can be either columnar, laminar, or isotropic, depending on the deposition conditions such as temperature, pressure, and composition of the input gases.[1][10][14][20] It is possible to obtain the desired structure by the proper control of these deposition parameters.

3.2   Columnar and Laminar Structures

Columnar Structure. The columnar structure of pyrolytic graphite is shown in Fig. 7.4. The crystallites are deposited with the basal planes (*ab* directions) essentially parallel to the deposition surface. Their structure tends to be columnar (cone-like) as a result of uninterrupted grain growth toward the reactant source.

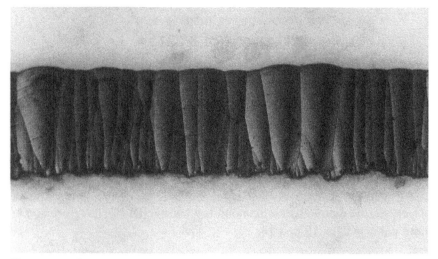

Figure 7.4. Columnar structure of pyrolytic graphite.

Effect of Substrate Geometry. The smoothness of the substrate is a very critical factor. Fig. 7.5 shows schematically the growth of the graphite deposit above a convex surface defect of the substrate such as a surface asperity or a dust particle. The deposit tends to magnify any such surface imperfection and, to obtain a uniform pyrolytic graphite growth, a perfectly smooth and clean surface is necessary.[11]

Continuous Nucleation. The structure is also often dependent on the thickness of the deposit. For instance, the grain size increases as the thickness increases and the columnar-grain structure becomes more pronounced as the film becomes thicker. Such large columnar structures are usually undesirable as the deleterious effects of grain growth and columnar formation can be considerable, causing structural failure and the rapid diffusion of impurities along grain boundaries.

Large grain size can be avoided by continuously adding growth sites, where new columnar growth can be generated. This effect is shown schematically in Fig. 7.6. These new growth sites originate from soot particles, which are formed in the gas-phase when the pressure and supersaturation are above a certain level.[11][14]

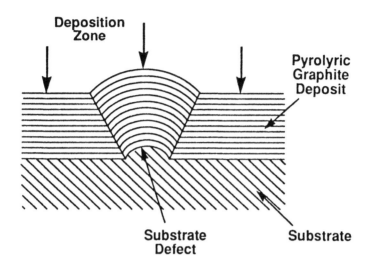

Figure 7.5. Effect of substrate defect on deposited structure of pyrolytic graphite.[11]

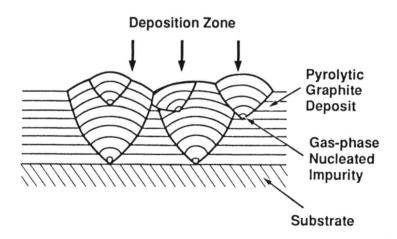

Figure 7.6. Effect of gas-phase nucleated impurities on deposited structure of pyrolytic graphite.[11]

Laminar Structure.   The laminar structure of pyrolytic graphite consists of essentially parallel layers (or concentric shells if deposited on a particle or fiber). It is shown in Fig. 7.7.

Both columnar and laminar structures are optically active to polarized light and usually have similar physical properties.[10][14]

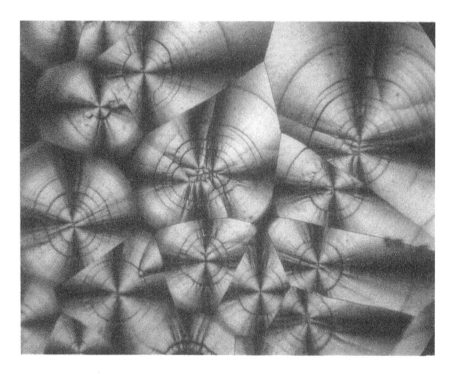

Figure 7.7. Laminar structure of pyrolytic graphite, deposited on carbon filament *(Photograph courtesy of Jack Chin, La Costa, CA.)*

3.3 Isotropic Structure

The other type of pyrolytic structure, isotropic carbon, has little graphitic characteristic and essentially no optical activity. It is composed of very fine grains without observable orientation and for this reason, it is known as isotropic carbon rather than isotropic graphite. It is often obtained in fluidized-bed deposition, possibly due to continuous surface regeneration by the mechanical rubbing action of the bed. An isotropic structure, observed by transmission electron microscopy, is shown in Fig. 7.8.[21]

Figure 7.8. High-density (~ 2.0 g/cm³) isotropic structure of pyrolytic carbon, observed by transmission electron microscopy. Viewing plane is parallel to deposition plane (x = 23,600). *(Photograph courtesy of J. L. Kaae, General Atomics, San Diego, CA.)*

3.4   Effect of Deposition Parameters

Effect of Pressure. Pyrolytic-graphite coatings with more uniformity, better coverage, and improved quality are generally obtained at low deposition pressure. Pressure controls the thickness of the surface boundary layer and consequently the degree of diffusion. By operating at low pressure, the diffusion process can be minimized and surface kinetics becomes rate-controlling.[3] Low-pressure deposition tends to be isotropic.

At higher pressure (i.e., atmospheric), the reactant gas must be diluted with an non-reactive gas such as hydrogen or argon to prevent uncontrolled vapor-phase precipitation, while generally no dilution is necessary at low pressure. However, atmospheric pressure reactors are simpler and cheaper, and, with proper control of the deposition parameters, satisfactory deposits can be obtained.

Effect of C/H Ratio. The C/H ratio of the gas mixture (CH_4, and H_2) entering the reaction chamber is an important factor in the control of the nature of the deposition. Higher C/H ratios (1/4) favor laminar deposition and lower ratios (1/14) favor isotropic deposition.[10]

Effect of Temperature. Generally, isotropic deposits are obtained at higher temperatures (>1400°C) and laminar and columnar deposits at lower temperatures.

In summary, isotropic deposits are obtained at high temperature, low pressures, and low carbon-to-hydrogen (C/H) ratio. The opposite conditions favor the deposition of laminar and columnar deposits.

3.5   Heat-Treatment and Graphitization

The mechanism of graphitization of pyrolytic graphite is essentially the same as that of pitch coke, described in Ch. 4, Sec. 3.3.[22]

Graphitization of Columnar and Laminar Deposits. The columnar and laminar deposits described above have generally a turbostratic structure in the as-deposited condition, with a large interlayer spacing (~ 0.344 nm) as revealed by x-ray diffraction. The material graphitizes readily when heat-treated at 2500°C for 90 minutes.

The 2500°C heat-treatment causes the reordering of the structure. The basal planes coalesce and become more parallel and closer together. The various crystallite imperfections such as vacancies, stacking faults, dislocations, and rotational disorders, tend to heal and disappear; the crystallite size (L_c) increases; the 002 line narrows considerably and becomes close to the position of the ideal graphite line as the interlayer spacing (d) decreases to approach that of the ideal graphite crystal (0.3354 nm). This observed reduction of the interlayer spacing is attributed in part to the removal of interstitial elements, mostly carbon.[22]

When columnar or laminar pyrolytic graphites are annealed above 2700°C, usually under a pressure of several atmospheres, further ordering and stress relieving of the structure occur within each plane and between planes. The material is known as "highly oriented pyrolytic graphite (HOPG)". It is soft and structurally close to the ideal graphite crystal with an angular spread of the c axes of crystallites of less than one degree.[23]

Graphitization of Isotropic Deposits. Unlike columnar and laminar pyrolytic deposits, isotropic carbon does not graphitize readily and is, in this respect, similar to vitreous carbon. Some reduction in the interlayer spacing

(d) is usually observed, but rarely does it decreases below 0.344 nm. The crystallite size (L_c) remains small.

4.0 PROPERTIES OF PYROLYTIC GRAPHITE

4.1 Properties of Columnar and Laminar Pyrolytic Graphites

The structure of both columnar and laminar pyrolytic graphites is close to that of the ideal graphite crystal (reviewed in Ch. 3, Secs. 3 and 4). These graphites have a high degree of preferred crystallite alignment particularly after heat-treatment, and their properties tend to be anisotropic. Melting point, sublimation point, heat of vaporization, entropy, enthalpy, specific heat, and chemical properties are similar to that of the single-crystal graphite but other properties may vary significantly (Ch. 3, Sec. 7).

The properties of pyrolytic graphite are summarized in Table 7.2. The values listed were collected from the manufacturer's data sheets.[24]-[27] The spread in value may represent slightly different materials and differences in the degree of graphitization. It may also reflect variations in the test methods; for instance, the measurement of mechanical properties may vary widely depending on the sample geometry and the test method (see Ch. 5, Sec. 2.1).

Table 7.2. Properties of Oriented Pyrolytic Graphite at 25°C

Density, g/cm³	2.10 - 2.24
Flexural strength tested in the c direction (across grain), MPa	80 - 170
Tensile strength tested in the ab directions (with grain), MPa	110
Young's modulus of elasticity, GPa	28 - 31
Thermal conductivity, W/m·K c direction ab directions	1 - 3 190 - 390
Thermal expansion 0 - 100°C, x10⁻⁶/m·K c direction ab directions	15 - 25 -1 to 1
Electrical resistivity, $\mu\Omega$·m c direction ab directions	1000 - 3000 4 - 5

Mechanical Properties.   The mechanical properties of pyrolytic graphite are like those of the ideal graphite crystal in the sense that they show a marked increase with increasing temperature as shown in Fig. 7.9.[11][26] Above 2600°C, the strength drops sharply.

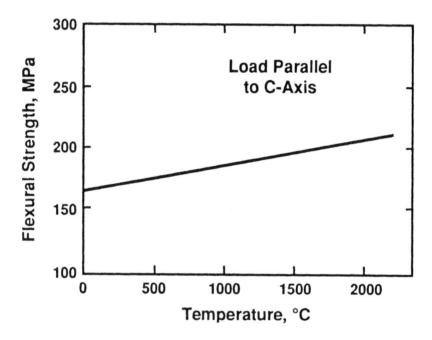

Figure 7.9. Flexural strength of pyrolytic graphite as a function of temperature.[11]

Thermal Conductivity.   As noted in Ch. 3, Sec. 4.3, the thermal conductivity of the ideal graphite crystal in the *ab* directions is high. It is far lower for the average pyrolytic graphite (up to 390 W/m·K at 25°C) but still high enough for the material to be considered a good thermal conductor, similar to copper (385 W/m·K at 25°C) (see Ch. 3, Table 3.6).

The thermal conductivity in the *c* direction is approximately 2.0 W/m·K at 25°C and, in this direction, graphite is a good thermal insulator, comparable to most plastics.   The anisotropy ratio is approximately 200.

The thermal conductivity in both the *ab* and *c* directions decreases with temperature as shown in Fig. 3.12 of Ch. 3.

Thermal Expansion. The thermal expansion of pyrolytic graphite, like that of the ideal graphite crystal, has a marked anisotropy. It is low in the *ab* directions (lower than most materials) but an order of magnitude higher in the *c* direction. The effect of temperature is shown in Fig. 7.10. Such a large anisotropy may lead to structural problems such as delamination between planes, especially in thick sections or when the material is deposited around a sharp bend.

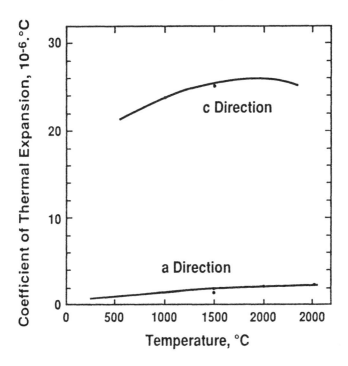

Figure 7.10. Coefficient of thermal expansion (CTE) of pyrolytic graphite as a function of temperature.[24]

Electrical Properties. The electrical properties of pyrolytic graphite also reflect the anisotropy of the material and there is a considerable difference between the resistivity in the *ab* and the *c* directions. Pyrolytic graphite is considered to be a good electrical conductor in the *ab* directions, and an insulator in the *c* direction. Its electrical resistivity varies with temperature as shown in Fig. 7.11.

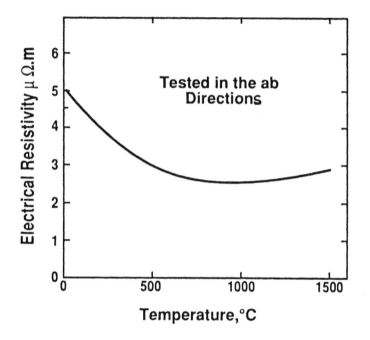

Figure 7.11. Electrical resistivity of pyrolytic graphite as a function of temperature.[26]

4.2 Properties of Isotropic Pyrolytic Carbon

As mentioned above, the structure of pyrolytic carbon is essentially isotropic and so are its properties. These are summarized in Table 7.3.

Hardness. Being composed of minute crystallites with essentially random orientation, isotropic pyrolytic carbon lacks the easy interlayer slippage which is characteristic of the well-oriented laminar or columnar structures of pyrolytic graphite. As a result, it is considerably harder. This makes it easy to polish and the material can be given a high gloss. The wear resistance is usually superior to that of the columnar and laminar deposits of vitreous carbon.[25]

Mechanical Properties. Isotropic pyrolytic carbon is stronger than the oriented pyrolytic graphites and vitreous carbon and can be considered as a reasonable structural material, comparing favorably with some of the more brittle ceramic materials.

Table 7.3. Properties of Isotropic Pyrolytic Carbon at 25°C (as deposited)

Density, g/cm^3	2.1
Vickers DPH* hardness, kg/mm^2	240 - 370
Flexural strength, MPa	350
Young's modulus, GPa	28
Strain to failure, %	1.2

* DPH: diamond pyramid hardness

Gas Permeability. Isotropic pyrolytic carbon provides a better barrier to gases than the more crystalline pyrolytic graphites.[26] The permeability, K, is as follows:

Isotropic pyrolytic carbon: 10^{-6} to 10^{-15} cm^2·s

Pyrolytic graphite: 10^{-2} to about 10 cm^2·s

5.0 APPLICATIONS OF PYROLYTIC GRAPHITE AND CARBON

5.1 High-temperature Containers and Other Free-Standing Products

Containers and other free-standing (monolithic) parts are produced by depositing the pyrolytic carbon or graphite onto a graphite mandrel. After the deposition is completed, the mandrel is removed, usually by machining.

It is difficult to deposit pyrolytic graphite in shapes having sharp radii without interlayer delaminations. These delaminations are caused by the stresses generated by the considerable differences in the thermal expansion in the *ab* and *c* directions which can overcome the low interlaminar strength. These delaminations do not occur with isotropic pyrolytic carbon.

Common applications of pyrolytic graphite are:

- Boats, and crucibles for *liquid-phase epitaxy*
- Crucibles for *molecular-beam epitaxy*
- Reaction vessels for the gas-phase epitaxy of *III-V semiconductor* materials such as gallium arsenide
- Trays for silicon-wafer handling

Free-standing isotropic pyrolytic carbon is a material of choice for solid propellant rocket nozzles. Because of its strength, hardness, and isotropic nature, it is able to withstand the mechanical erosion which is the dominant failure mechanism above 3200°C.[27]

5.2 Resistance-Heating Elements

Because of its good electrical characteristics and refractoriness, pyrolytic graphite is used extensively for high-temperature resistance-heating elements. In combination with pyrolytic boron nitride (PBN), it provides an integrated heating system in which the graphite is the resistive element and PBN the insulating substrate. Both materials are produced by CVD. The product is used as a source heater in metal evaporation and semiconductor epitaxy, as a substrate heater in thin-film deposition, as a melt heater for crystal growth, and in other applications.[28]

5.3 Nuclear Applications

Isotropic pyrolytic carbon exhibits excellent stability under neutron irradiation. This, coupled with its high strength, dense isotropic structure, and impermeability to gases, makes it the material of choice for the coating of nuclear fission particles to contain the fission products.[1][24] The coating is produced in a fluidized bed (see Sec. 2.7 above).

5.4 Biomedical Applications

Biomedical applications require a material with good strength, fatigue-resistance, high erosion resistance, chemical inertness, and compatibility with blood and tissues. Isotropic pyrolytic carbon meets these criteria and is used extensively in biomedical devices such as heart valves and dental implants where its performance is superior to other forms of carbon such as pyrolytic graphite or vitreous carbon.[25]

5.5 Coatings for Molded Graphites

A pyrolytic graphite coating, applied to a molded graphite substrate, provides a smooth, essentially pore-free surface that can enhance the chemical resistance. Such coated parts are found in applications requiring chemical inertness at high temperature (in a non-oxidizing atmosphere) such as the following:

- Wafer trays for plasma-CVD equipment
- Boats for *liquid-phase epitaxy*
- Boats, and other parts for vapor deposition of *III-V semiconductor* compounds such as gallium arsenide
- Hardware for metal processing

5.6 Coatings for Fibers

Inorganic fibers, such as silicon carbide or alumina, provide the reinforcement for metal or ceramic matrices to form refractory composites. These fibers often react with the matrix material during processing into composites or during operation at high temperature. This interaction produces intermetallics or other compounds which may considerably degrade the properties of the composite. A pyrolytic graphite coating applied on the fiber acts as a diffusion barrier, and prevents these diffusional reactions (see Ch. 9, Secs. 5.0 and 6.0).[3][29]

Another fiber coating application is found in fiber optics. An isotropic pyrolytic-carbon coating is applied on optical fibers to improve the abrasion and fatigue resistance, and bending performance.[30]

5.7 Carbon-Carbon Infiltration

Pyrolytic carbon or graphite is used to densify carbon-carbon structures by infiltration as described in Sec. 2.6 above. Applications include reentry heat shields, rocket nozzles, aircraft disk brakes, and other aerospace components.[17] This topic is reviewed in more detail in Ch. 9.

REFERENCES

1. Powell, C. F., Oxley, J. H., and Blocher, J. M. Jr., *Vapor Deposition*, John Wiley and Sons, New York (1966)

2. Sawyer, W. E. and Man, A., *US Patent* 229335 (June 29, 1880)

3. Pierson, H. O., *Handbook of Chemical Vapor Deposition*, Noyes Publications, Park Ridge, NJ (1992)

4. Picreaux, S. and Pope, L., *Science*, 226:615-622 (1986)

5. Lucas, P. and Marchand, A., *Carbon*, 28(1):207-219 (1990)

6. Chase, M. W., *JANAF Thermochemical Tables*, Vol. 13, Supp. No. 1, American Chem. Soc. & Am. Inst. of Physics (1985)

7. Wagman, D. D., The NBS Tables of Chemical Thermodynamic Properties, *J. Phys. Chem. Ref. Data 11*, and supplements (1982); Barin, I. and Knacke, O., *Thermochemical Properties of Inorganic Substances*, Springer, Berlin (1983); Hultgren, R., *Selected Values of the Thermodynamic Properties of the Elements*, Am. Soc. for Metals, Metals Park, OH (1973)

8. Besmann, T. M., *SOLGASMIX-PV, A Computer Program to Calculate Equilibrium Relationship in Complex Chemical Systems*, ORNL/TM-5775, Oak Ridge National Laboratory, Oak Ridge TN (1977)

9. EKVICALC, and EKVIBASE, *Svensk Energi Data*, Agersta, S-740 22 Balinge, Sweden

10. Pierson, H. O. amd Lieberman, M. L., *Carbon*, 13:159-166 (1975)

11. Campbell, I. and Sherwood, E. M., *High Temperature Materials and Technology*, John Wiley & Son, New York (1967)

12. Motojima, S., Kawaguchi, M., Nozaki, K. and Iwanaga, H., *Proc. 11th. Int. Conf. on CVD*, (K. Spear and G. Cullen, eds.), 573-579, Electrochem. Soc., Pennington, NJ 08534 (1990)

13. Gower, R. P. and Hill, J., *Proc. 5th Int. Conf. on CVD*, (J. Blocher and H. Hintermann, eds.), 114-129, Electrochem. Soc., Pennington, NJ 08534 (1975)

14. Bokros, J. C., *Chemistry and Physics of Carbon*, Vol. 5 (P. L. Walker, ed.), Marcel Dekker Inc., New York (1969)

15. Lackey, W. J., *Ceram. Eng. Sci. Proc.*, 10(7-8):577-584 (1989)

16. Starr, T. L., *Ceram. Eng. Sci. Proc.*, 9(7-8):803-812 (1988)

17. Buckley, J. D., *Ceramic Bulletin*, 67(2):364-368 (1988)

18. Kaae, J. L., *Ceram. Eng. Sci. Proc.*, 9(9-10):1159-1168 (1988)

19. Inspektor, A., Carmi, U., Raveh, A., Khait, Y. and Avni, R., *J. Vac. Sci. Techn.*, A 4 (3):375-378 (1986)

20. Lieberman, M. L. and Pierson, H. O., *Proceeding of the 11th. Carbon Conference* (June 1973)

21. Kaae, J. L., *Carbon*, 223(6):665-673 (1985)

22. Kawamura, K. and Gragg, R. H., *Carbon*, 24(3):301-309 (1986)

23. Kavanagh, A. and Schlogl, R., *Carbon*, 26(1):23-32 (1988)

24. *Pyro-tech PT-101,* Technical Brochure, Ultra Carbon Corp., Bay City, MI 48707

25. Beavan, A., *Materials Engineering*, 39-41 (Feb. 1990)

26. *Graphite, Refractory Material,* Bulletin from Le Carbone-Lorraine, Gennevilliers 92231, France

27. *Products for the Semiconductor Industry,* Bulletin from Ringsdorff, D-5300 Bonn-2, Germany (1988)

28. *Boralectric,* Bulletin from Union Carbide Coatings Corp., Cleveland OH, 44101-4924 (1989)

29. Cranmer, D. C., *Ceramic Bulletin*, 68(2):415-419 (1989)

30. Morrow, A., *Laser Focus World*, 189-190 (January 1989)

8

Carbon Fibers

1.0 GENERAL CONSIDERATIONS

1.1 Historical Perspective

Thomas Edison made the first carbon fibers on record in 1879 when he carbonized cotton thread to produce a filament for a light bulb. His effort was not entirely successful and Edison eventually replaced the fiber by a tungsten wire. Large-scale production of carbon fibers had to wait until the late 1950's, when cloth and felt from carbonized rayon were commercialized. These materials are still produced now (see Sec. 4.0).

High-strength fibers from polyacrylonitrile (PAN) were developed in the 1960's by workers in Japan and in the UK, and pitch-based fibers saw their start in Japan and the U.S., at about the same time.[1]-[3]

1.2 The Carbon-Fiber Business

Fibers, whether natural or synthetic, occupy a large place in the industrial world. They are usually classified according to their origin as organic and inorganic. The organic (or textile) fibers are available in many varieties such as cotton, wool, rayon, and many others. They form a huge but somewhat static market nearing $10 billion in the U.S. alone in 1987.

166

Inorganic fibers which include structural glass, ceramics, boron, and carbon, form a smaller but growing market estimated at $1.8 billion worldwide in 1990.[4][5] Optical glass, which is a specialized product, is not included. The inorganic fiber market is divided as shown in Table 8.1.

Table 8.1. Estimated World Inorganic-Fiber Production and Cost in 1991

Fiber	Production (metric ton)	Average Cost ($/kg)	World Market ($ million)
Glass	800,000	2	1,600
Boron	15	800	12
Ceramics	120	90	11
Carbon	5,000	40	200
		Total	1,823

Glass is still by far the dominant fiber material, especially on a tonnage basis, but it is increasingly challenged by carbon in many high-performance applications, and the production of carbon fibers has increased rapidly in the last 20 years, going from a few tons per year to the present level estimated at 5000 tons.

This rapid growth is the direct result of two major factors:

1. The remarkable properties of these fibers, properties which are continuously upgraded

2. The gradual decrease in cost due to improvements in the manufacturing process. As shown in Fig.8.1 for PAN-based fibers, the cost decrease in 20 years is 80% (90% when adjusted for inflation).[6]

It is likely that these two trends, improvement of the properties and cost reduction, will continue for some time and that production and the number of applications of carbon fibers will keep on expanding.

This expansion of the market is fueled by the rapid development of advanced-composite materials where carbon fibers now play a predominant role. These composites include reinforced plastics, carbon-carbon, and metal and ceramic composites. They are found in aerospace, defense, automotive, sporting goods, and other areas (see Ch. 9).

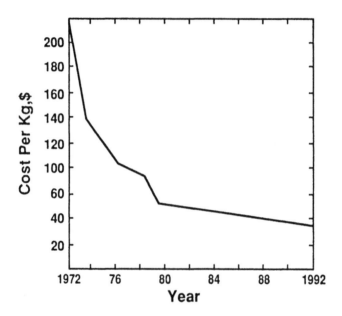

Figure 8.1. Cost of PAN-based carbon fibers (not adjusted for inflation).

While the potential of carbon fibers is impressive, it has been the object at times of over-optimistic predictions, some of which were believed. This in turn led to a overbuilding of production facilities and excess of production capacity. In 1992, the worldwide production was estimated at five million tons while the production capacity (as opposed to actual production) stood well above that figure. The business is obviously attractive but extremely competitive.

Japanese Dominance and Internationalization. The carbon-fiber business is dominated at the present by Japanese companies such as Toray, Toho, Mitsubishi Rayon, Sumika, and Asahi Nippon. These firms, to a great degree, control the patents, the licensing, and the manufacturing of carbon fibers produced from the basic wet-spun PAN precursor. This precursor is also produced by the British firm Courtauld but is not produced in the U.S.[7]

The carbon-fiber industry is going through rapid changes and, at times, bewildering gyrations, both in the U.S. and abroad, as companies change ownership, licenses are granted, and partnerships created and dissolved. In addition, the internationalization of the business is spreading and it is becoming increasingly difficult to group the producers of carbon fibers by country since many operations are international in character and ownership.

Table 8.2 is a partial list of carbon-fiber producers in operation in 1992, with their trademark and licences.

Table 8.2. Producers of Carbon Fibers in 1992, Partial List
(Trademark in brackets)

Akzo, Germany and Knoxville, Tn, Toho Rayon Licensee (Tenax)

Amoco Performance Products, Atlanta, GA, Toray licensee (Thornel)

Ashland Carbon Fibers, Inc., Ashland, KY (Carbofles)

BASF, Germany, and Charlotte, NC, Toho Licensee (Celion)

BP Chemicals, UK, and Santa Ana, CA

Enka, Germany, Toho licensee (Carbolon)

Fortafil Carbon, Rockwood, TN, subsidiary of Akzo (Fortafil)

Grafil, Sacramento, CA, subsidiary of Mitsubishi Rayon (Grafil)

Hercules, Magma, UT (Hercules)

Kureha Chemicals, Japan

Mitsubishi Rayon, Japan (Pyrofil)

Morganite, UK (Modmor)

Nippon Petrochemicals, Japan (Granoc)

Rolls-Royce, UK

Sigri Elektrographit, Germany (Sigrafil)

Soficar, France, Toray licensee

Toho Rayon Co., Japan (Bestfight)

Toray, Japan, and Kirkland, WA (Torayca)

1.3 Carbon and Graphite Nomenclature

As seen in Chs. 3 and 4, carbon and graphite materials are polycrystalline aggregates, composed of small graphitic crystallites with varying degree of basal plane alignment.

The structure of carbon fibers is mostly turbostratic with little graphitic character and may in some cases include sp^3 bonds (see below). For that reason, these fibers are usually and generally more accurately called "carbon fibers". The terminology "graphite fiber" is still found in many references in the literature and certainly could be justified in some cases such as the highly-oriented pitch-based fibers. However, to simplify matters, "carbon fibers" will be used in this book.

1.4   Competing Inorganic Fibers

In the competing world of today, where performance and economics are the keys to success, it is necessary to compare carbon fibers with other inorganic-fiber contenders in order to obtain the proper perspective.

These competitors are produced from glass, ceramic oxides, silicon carbide, and boron. Table 8.3 compares their main characteristics and lists typical applications.

Table 8.3. Characteristics and Applications of Inorganic Fibers

Material	Main Characteristics	Typical Applications
Glass	High strength Low modulus Low temperature Low cost	Insulation Reinforced plastics
Oxides	Medium strength Good oxidation resistance High cost	Ceramic composites High-temp. insulation Filtration
Silicon carbide	High strength High modulus High cost	High-temp. composites
Boron	High strength High modulus High cost	Plastic and metal composites
Carbon	High strength High modulus Low density Low oxidation resistance Medium cost	Reinforced plastics Carbon-carbon High-temp insulation

The tensile properties and densities of typical inorganic fibers are shown in Table 8.4 (data are supplied by the manufacturers). Carbon fibers compare favorably overall with other fibers.

Table 8.4. Tensile Strength, Modulus, and Density of Selected Inorganic Fibers

Major Constituent	Strength (MPa)	Tensile Modulus (GPa)	Tensile Density (g/cm³)
Alumina (1)	1750	154	2.7
Alumina (2)	2275	224	3.0
Silicon carbide (3)	3920	406	3.0
Boron (4)	3600	400	2.5
Glass (5)	4580	86	2.5
Carbon (6)	5500	330	1.7

(1) Nextel 312, 3M, Minneapolis, MN
(2) Nextel 400
(3) Avco SCS 6, Textron Inc. Lowell, MA
(4) Avco Boron
(5) S-Glass, Corning Glass, Corning, NY
(6) MS-40, Grafil, Sacramento, CA

Specific Strength Properties. Specific strength and modulus (that is the strength and modulus values divided by the density) are important in applications where weight is critical such as in aerospace. Carbon fibers have high specific strength because of their low density and are clearly at an advantage over other high-strength fibers as shown in Fig.8.2.[7][8]

Glass Fibers.[9] Structural glass fibers are the oldest and the most-widely used inorganic fibers. They account for the great majority of the fibers found in reinforced plastics and in fibrous insulation. The production process is simple and low cost. Theses fibers have high strength and low cost but also low modulus and poor resistance to high temperature.

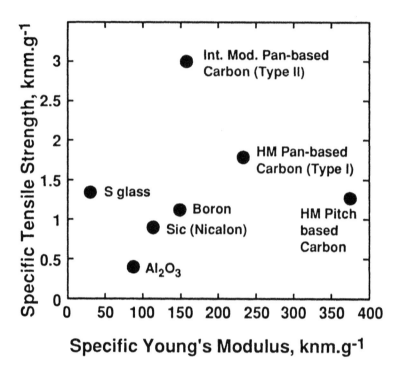

Figure 8.2. Specific strength and modulus of carbon and other fibers (average values from manufacturers' data).

Ceramic Fibers. The major use of ceramic fibers was, until recently, in high-temperature insulation. Recent improvements in manufacturing technology have opened new applications, particularly in advanced composites, insulation, and chemical components. The main advantage of these fibers is refractoriness in general and, for oxide fibers, good oxidation resistance and strength retention up to 1200°C. From that standpoint, oxide fibers are considerably better than carbon fibers. Most ceramic fibers are still expensive.

Boron and Silicon-Carbide Fibers. Boron and silicon-carbide (SiC) fibers have high strength and modulus, but have low oxidation resistance and high cost. They are produced by chemical-vapor deposition (CVD) and sol-gel methods.

1.5   State of the Art

Carbon fibers currently in production can be divided into three categories based on the precursor materials: PAN (polyacrylonitrile), pitch, and rayon, each with its own characteristics and special advantages.

- PAN-based fibers constitute the largest segment of the industry. These fibers have high strength and modulus and, as mentioned above, their cost is continually decreasing.

- Pitch-based fibers can be divided in two groups: *(a)* the isotropic pitch fibers, which have low mechanical properties and relatively low cost, and *(b)* the mesophase-pitch fibers, which have very high modulus but are more expensive. Pitch-based fibers are produced by a simple process and their cost should eventually rival that of glass fibers. Their potential is yet to be fully realized.

- Rayon-based fibers are not as strong as PAN-based fibers. They are used in insulation and some carbon-carbon and ablative applications because of a good match of properties with the carbonized matrix.

In addition to these, several other categories are being developed. These include the vapor-phase (CVD) fibers which offer promise for the development of whiskers, and fibers based on other precursors, such as polyvinyl chlorides, Saran™ and polyphenylene benzobisthiazole.[10] These fibers are still experimental and have yet to reach the production stage.

2.0   CARBON FIBERS FROM PAN

2.1   PAN as Precursor

Polyacrylonitrile (PAN) is the most favored precursor at this time for the production of high-performance fibers. PAN-based fibers are in the medium price range (~ \$40/ kg in 1992). They have the highest tensile strength of all carbon fibers and a wide range of moduli and are readily available in a variety of *tows*.

Precursor Requirements. To produce the optimum carbon fiber, a precursor fiber should perform the following functions:[5]

- Maximize the preferred orientation of the polycarbon layers in the direction of the fiber axis to achieve highest stiffness.

- Preserve the highest degree of defects in order to hinder the formation of crystalline graphite with its low shear and modulus.

- Preserve the fibrous shape of the polymer precursor during pyrolysis.

- Provide as high a carbon yield as possible.

PAN essentially meets the requirements. It was originally developed as an organic synthetic textile fiber and this is still by far its major use. It has also shown to be an ideal precursor for the production of carbon fibers, but only when produced by the wet-spinning process (as opposed to the more common and cheaper dry-spinning).

PAN is usually co-polymerized with a small amount of another monomer such as methyl acrylate or vinyl acetate. This co-polymerization lowers the glass-transition temperature from 120°C to <100°C and, as a result, improves the spinning and stretching characteristics.

2.2 Processing of PAN-based Carbon Fibers

The production process of PAN-based carbon fibers can be divided into the following five steps:[11]

1. Spinning the PAN co-polymer to form a fiber

2. Stretching

3. Oxidation and stabilization at 220°C under tension

4. Carbonization in inert atmosphere up to 1600°C

5. Graphitization up to 3000°C

Spinning. Since PAN decomposes before melting, it is necessary to make a solution with a solvent such as dimethyl formamide in order to be able to spin the material into a fiber. The spinning operation is either dry spinning, where the solvent evaporates in a spinning chamber (Fig. 8.3 a) or wet spinning, where the fiber is placed in a coagulating bath solution

(Fig. 8.3 b).[5] In the first case, the rate of solvent removal is greater than the rate of diffusion within the fiber and the surface of the filament hardens faster than the interior, resulting in a collapse and the formation of a dog-bone cross section.[12] In the second case, the fiber dries uniformly and the cross section is circular.

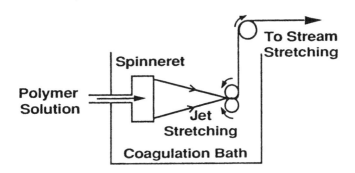

a) Wet Spinning (320 000 Filaments)

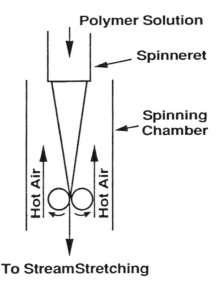

b) Dry Spinning (2500 Filaments)

Figure 8.3. Schematic of spinning processes of PAN fibers.[5]

Of the two fibers, only the wet-spun PAN is used as precursor. It contains a co-polymer, such as itaconic acid or other proprietary compounds, that apparently catalyzes the cyclization in air and helps the carbonization process.[7] The dry-spun fiber is not as suitable and is not used.

Stretching. The spun fiber is composed of a fibrillar or ribbon-like network, which acquires a preferred orientation parallel to the fiber axis, providing that the fiber is stretched either while it is still in the coagulating bath, or subsequently in boiling water, as shown in Fig. 8.4. This stretching results in an elongation of 500 to 1300%, and is an essential step to obtain a high-strength fiber.

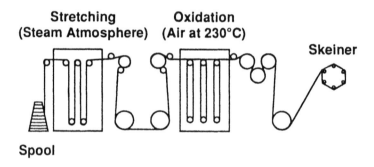

Figure 8.4. Schematic of stretching and oxidation steps in the production of PAN-based carbon fibers.[5]

Stabilization and Oxidation. During the carbonization process, the elimination of the non-carbon elements (hydrogen and nitrogen) is usually accompanied by chain scission and relaxation of the fibrillar structure. This is detrimental to the formation of high-strength and high-modulus fibers, but can be avoided by a stabilization process prior to carbonization.

This stabilization consists of slowly heating the stretched fiber to 200 - 280°C in an oxygen atmosphere (usually air) under tension to maintain the orientation of the polymer skeleton and stabilize the structure (Fig. 8.4). The addition of ammonia to oxygen increases the rate of stabilization.[13]

The molecular changes occurring during stabilization are shown in Fig.8.5.[11] The oxidation causes the formation of C=C bonds and the incorporation of hydroxyl (-OH) and carbonyl (-CO) groups in the structure. These factors promote cross-linking and thermal stabilization of the fiber which, at this stage, can no longer melt. This non-meltable characteristic is essential to prevent the filaments from fusing together.

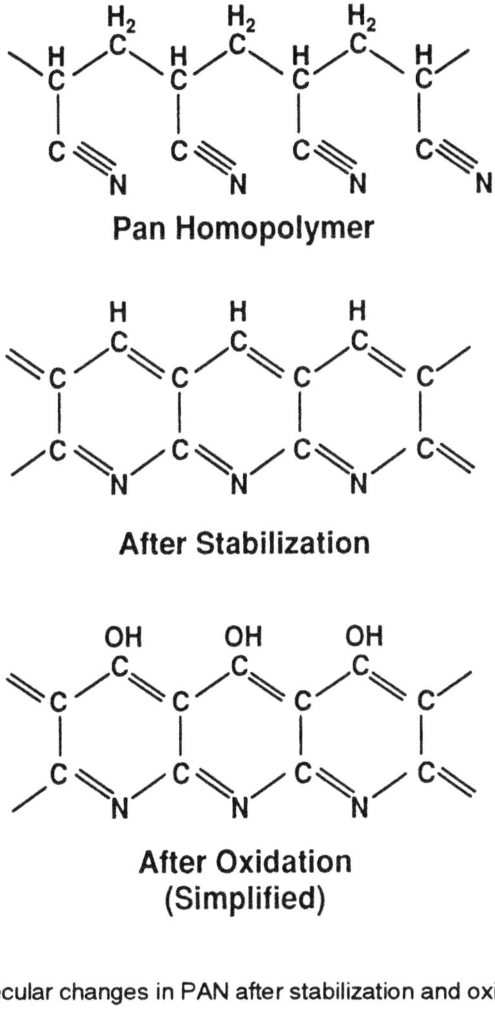

Pan Homopolymer

After Stabilization

After Oxidation
(Simplified)

Figure 8.5. Molecular changes in PAN after stabilization and oxidation.[11]

Carbonization and Graphitization. Carbonization takes place be-tween 1000 and 1500°C. These temperatures are reached slowly, at a heating rate of ~ 20°C/min. During this stage a considerable amount of volatile by-products is released. These include H_2O, CO_2, CO, NH_3, HCN, CH_4, and other hydrocarbons. The carbon yield is between 50 and 55%. The circular morphology of the fiber is maintained and the final diameter varies from 5 to 10 μm, which is approximately half that of the precursor PAN fiber. The removal of nitrogen occurs gradually over a range of temperatures as shown below:[11]

600°C – nitrogen evolution starts

900°C – maximum evolution

1000°C – 6% nitrogen left

1300°C – 0.3% nitrogen left

The tensile modulus of the fiber can be further increased by graphiti-zation. It can be argued that the term "graphitization" is not correct since a true graphite structure is not obtained, and "high-temperature heat-treat-ment" would be a better term. This heat-treatment is usually carried out at temperatures up to 2500°C. The final carbon content is greater than 99%.

2.3 Structure of PAN-based Carbon Fibers

Analytical Techniques. Analytical techniques to determine the structure of carbon fibers include: wide-angle and small-angle x-ray diffrac-tion, electron diffraction, neutron scattering, Raman spectroscopy, electron microscopy, and optical microscopy. Detailed reviews of these techniques are found in the literature.[14]

Structure. The structure of PAN-based carbon fibers is still conjec-tural to some degree. Yet, thanks to the recent advances in analytical techniques just mentioned, an accurate picture is beginning to emerge.

Unlike the well-ordered parallel planes of pyrolytic graphite which closely match the structure of the graphite crystal, the structure of PAN-based carbon fibers is essentially turbostratic and is composed of small two-dimensional fibrils or ribbons. These are already present in the precursor and are preferentially aligned parallel to the axis of the fiber. The structure may also include lamellas (small, flat plates) and is probably a combination of both fibrils and lamellas.[1][15]

Crystallite Size. Several structural models have been proposed including the one shown in Fig.8.6.[11] The critical parameters (as determined by x-ray diffraction) are L_c, which represents the stack height of the ribbon and the crystallite size L_a, which in this case can be considered as the mean length of a straight section of the fibril.[11] This alignment (L_a) becomes more pronounced after high-temperature heat-treatment which tends to straighten the fibrils. However, L_a still remains small and is generally less than 20 nm, as shown in Fig.8.7.[7] This figure also shows the much greater increase of crystallite size (La) of pitch-based fibers (see Sec. 3.0 below).

The straightening of the fibrils occurs preferentially, the outer fibrils being more oriented (straightened) than the inner ones as shown in Fig. 8.8.[11] This has an important and favorable consequence, that is, most of the load-bearing capacity is now transferred to the outer portion or "skin" of the fiber.

Interlayer Spacing. The change in interlayer spacing (c spacing) of PAN-based carbon fibers as a function of heat-treatment temperature is shown in Fig.8.9.[7] This spacing never shrinks to less than 0.344 nm, even after a 3000°C heat-treatment, indicating a poor alignment of the basal planes and the presence of defects, stacking faults, and dislocations. This behavior is characteristic of carbons produced from polymers (see Ch. 6). Also shown in Fig. 8.9 is the decrease in interlayer spacing of a pitch-based fiber. It is far more pronounced than that of the PAN-based fiber (see Sec. 3.0 below).

Sp^2 and Sp^3 Bonding. Another important structural characteristic of PAN-based fibers is the probable existence of sp^3 hybrid bonding as indicated by Raman spectroscopy and shown in Fig. 8.10. In this figure, the pitch-based graphitized fiber (P100) is the only one to exhibit a strong sp^2 line. All others show structural disorders which may be caused by some sp^3 bonding.[7] The fibers listed in Fig. 8.10 are identified in Secs. 6.3 and 6.4 below.

Both sp^2 and sp^3 hybrid bonds are strong covalent bonds, with the following bond energy and bond lengths (see Ch. 2, Secs. 3 and 4):

$$sp^3 \quad - \quad 370 \text{ kJ/mol and } 0.15 \text{ nm}$$

$$sp^2 \quad - \quad 680 \text{ kJ/mol and } 0.13 \text{ nm}$$

These strong bonds within the crystallites (or fibrils) and the preferred orientation of these crystallites account, at least in part, for the high stiffness inherent to most carbon fibers.

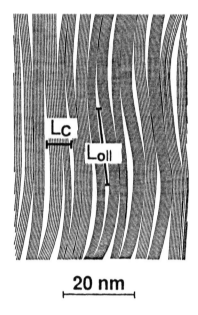

Figure 8.6. Proposed structural model for carbon fiber.[11]

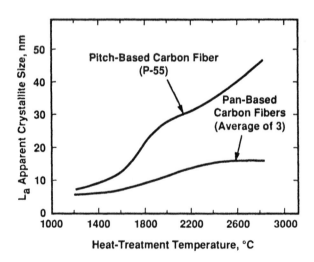

Figure 8.7. Apparent crystalline size (L_a) of PAN-based and pitch-based carbon fibers as a function of heat-treatment temperature.[7]

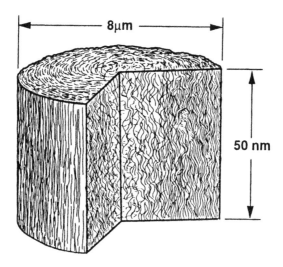

Figure 8.8. Model of a PAN-based carbon fiber in cross-section showing the "skin" effect (not to scale).[11]

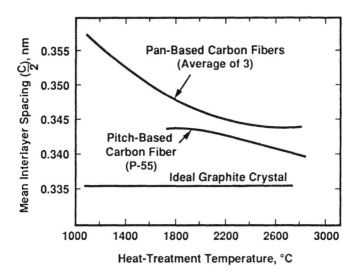

Figure 8.9. Mean interlayer spacing (c/2) of PAN-based and pitch-based carbon fibers as a function of heat-treatment temperature.[7]

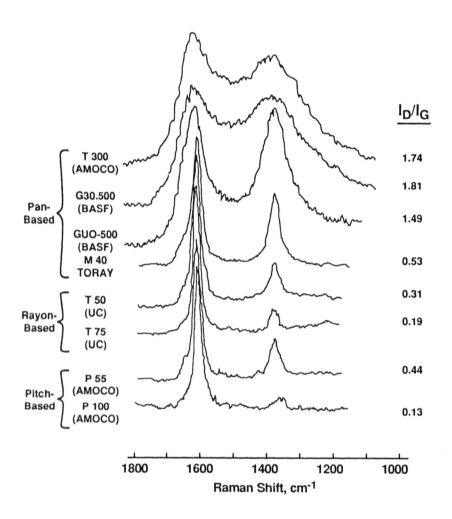

Figure 8.10. Laser Raman spectroscopy of PAN-based, rayon-based, and pitch-based carbon fibers.[7]

To summarize, small crystallite size, high interlayer spacing, and general structural disorder are the factors that contribute to the unique and stable turbostratic structure of PAN-based carbon fibers (which is likely to include both sp^2 and sp^3 hybrid bonds), and explain their inability to form a graphitic structure even after high-temperature heat-treatment (i.e., 3000°C).

3.0 CARBON FIBERS FROM PITCH

3.1 Pitch Composition

As seen in Ch. 5, Sec. 2.1, pitch is a by-product of the destructive distillation of coal, crude oil, or asphalt. It meets the conditions for carbon-fiber production stated above (Sec. 2.1) and is inexpensive and readily available. Its carbon yield can exceed 60%, which is appreciably higher than the yield of PAN (~ 50%).

The composition of pitch includes four generic fractions with variable proportions.[11] These are:

- The saturates, which are aliphatic compounds with low molecular weight similar to wax

- The napthene aromatics, which are low-molecular-weight compounds with saturated ring

- The polar aromatics, which are medium-molecular-weight compounds with some heterocyclic molecules

- The asphaltenes, which have high molecular weight and a high degree of aromaticity. The higher the ratio of asphaltene, the higher the softening point, thermal stability, and carbon yield.

3.2 Carbon Fibers from Isotropic Pitch

Low-cost carbon fibers are produced from an isotropic pitch with a low-softening point. The precursor is melt-spun, thermoset at relatively low temperature, and carbonized. The resulting fibers generally have low strength and modulus (~ 35 - 70 GPa). They are suitable for insulation and filler applications. Their cost dropped to less than $20/kg in 1992.[6]

3.3 Carbon Fibers from Mesophase Pitch

Carbon fibers from mesophase pitch have high modulus and medium strength. They are presently more costly than PAN-based fibers ($65/kg in 1992).

Precursor Pitch. The precursor material is a mesophase pitch, characterized by a high percentage of asphaltene. Table 8.5 shows the approximate composition of three common mesophase compounds.

Table 8.5. Composition of Mesophase Pitch

Product	Composition, %	
Ashland 240*	Asphaltene	62
	Polar aromatics	9
	Naphthene aromatics	29
Ashland 260*	Asphaltene	83
	Polar aromatics	7
	Naphthene aromatics	10
CTP 240**	Asphaltene	68
	Polar aromatics	11
	Naphthene aromatics	21

* Petroleum pitch, product of Ashland, Petroleum, Ashland, KY

**Coal tar pitch

Processing. The processing of mesophase-pitch fibers is similar to that of PAN fibers, except that the costly stretching step during heat-treatment is not necessary, making the process potentially less expensive.[16][17] The processing steps can be summarized as follows and are represented schematically in Fig. 8.11:

1. Polymerization of the isotropic pitch to produce mesophase pitch.

2. Spinning the mesophase pitch to obtain a "green fiber".

3. Thermosetting the green fiber.

4. Carbonization and graphitization of the thermoset fiber to obtain a high-modulus carbon fiber.

Polymerization. The pitch is heated to approximately $400^\circ C$ and is transformed from an isotropic to a mesophase (or liquid crystal) structure consisting of large polyaromatic molecules with oriented layers in parallel stacking. This structure is similar to the needle-coke stage of molded carbons described in Ch. 4, Sec. 2.3.

Pitch Polymerization

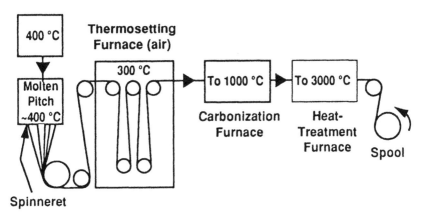

Figure 8.11. Schematic of the production steps in the manufacture of pitch-based carbon fibers.

Spinning and Thermosetting.[12] The mesophase pitch is melt-spun in a mono- or multifilament spinneret, heated to 300 - 450°C and pressurized with inert gas. It is drawn at a speed >120 m/min for a draw ratio of approximately 1000/1 to a diameter of 10 - 15 μm. The draw ratio is an important factor in the control of the orientation of the fiber structure: the higher the draw, the greater the orientation and uniformity.

At this stage the fiber is thermoplastic and a thermosetting operation is needed to avoid the relaxation of the structure and prevent the filaments from fusing together. This thermosetting operation is carried out in an oxygen atmosphere or in an oxidizing liquid at approximately 300°C, causing oxidation cross-linking and stabilization of the filament. Temperature control during this thermosetting step is critical since a temperature which is too high would relax the material and eliminate its oriented structure.

Carbonization and Heat-Treatment. The thermoset fibers are then carbonized at temperatures up to 1000°C. This is done slowly to prevent rapid gas evolution and the formation of bubbles and other flaws. Carbonization is followed by a heat-treatment from 1200 to 3000°C, at the end of which the final structure, strength, and modulus are established (see Sec. 6.4 below).[10][17]

3.4   Structure of Mesophase-Pitch Carbon Fibers

Cross Section. The cross-section structure of mesophase-pitch carbon fibers is one of the four types shown in Fig. 8.12 and is determined by the spinning method, the temperature of stabilization, and the partial pressure of oxygen.[2][19] The formation of a skin-core structure or skin effect is often observed. This structure is similar to that of the PAN-based carbon fiber shown in Fig. 8.8 above.

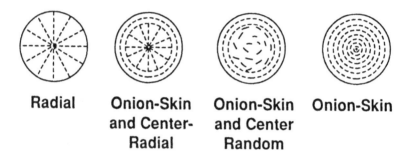

Radial   Onion-Skin   Onion-Skin   Onion-Skin
            and Center-   and Center
            Radial   Random

Figure 8.12. Cross-section of the various structures observed in pitch-based carbon fibers.[2]

Interlayer Spacing and Crystallite Size. The increase of the apparent crystallite size of mesophase-pitch fibers as a function of heat-treatment is shown in Fig. 8.7, and the decrease of the interlayer spacing (c spacing) in Fig.8.9.[7] The spacing of the heat-treated fiber is relatively close to that of the ideal graphite crystal (~ 0.340 nm vs. 0.3355 nm), indicating a decreased turbostratic stacking of the basal planes and a well-ordered structure. In this respect, the difference between these fibers and the PAN-based fibers is pronounced.

The large crystallites of the heat-treated pitch-based fibers, which is structurally close to the perfect graphite crystal and well aligned along the fiber axis, offer few scattering sites for phonons. This means that these fibers have a high thermal conductivity along the fiber axis since, as mentioned in Ch. 3, Sec. 4.3, the transfer of heat in a graphite crystal occurs mostly by lattice vibration (see Sec. 6.6 below).

However, this high degree of crystallinity also results in low shear and compressive strengths. In addition, these carbon fibers tend to have flaws such as pits, scratches, striations, and flutes. These flaws are detrimental to tensile properties but do not essentially affect the modulus and the thermal conductivity.[16]

4.0 CARBON FIBERS FROM RAYON

Rayon-based fibers were the first carbon fibers produced commercially. They were developed in the 1960's specifically for the reinforcement of ablative components for rockets and missiles. However, they are difficult to process into high-strength, high-modulus fibers and have been replaced in most structural applications by PAN or pitch-based fibers.

4.1 Rayon Precursor

A number of rayon fibers are available, but the most suitable is the highly polymerized viscose rayon. The molecular structure is as follows:[11]

$$
\begin{array}{cc}
\text{OH} & \text{OH} \\
| & | \\
\text{CH}_2 & \text{CH}_2 \\
| & | \\
\text{CH-O} & \text{CH-O} \\
\diagup \quad \diagdown & \diagup \quad \diagdown \\
\text{-CH} \qquad \text{CH-O-CH} & \text{CH-O} \\
\diagdown \quad \diagup & \diagdown \quad \diagup \\
\text{CH - CH} & \text{CH - CH} \\
| \quad | & | \quad | \\
\text{OH} \ \ \text{OH} & \text{OH} \ \ \text{OH}
\end{array}
$$

As can be seen, this structure has many heteroatoms (O and H) which must be removed. Moreover, many carbon atoms are lost due to the formation of volatile carbon oxides during pyrolysis. As a result the carbon yield is low (<30%) and shrinkage is high.

4.2 Processing

The rayon precursor is first heated to 400°C at the relatively slow rate of 10°C per hour. During this step, the fiber is stabilized; H_2O is formed from the hydroxyl groups in the molecule and the fiber depolymerizes with the evolution of CO and CO_2 and the formation of volatile, tar-like compounds. This depolymerization makes it impossible to stretch the fiber at this stage.

The heating rate can be increased and the carbon yield maximized by heating the fiber in a reactive atmosphere such as air or chlorine, or impregnating it with flame retardants and carbonization promoters.[11]

Carbonization is the next step and is carried out in an inert atmosphere at a temperature range of 1000 - 1500°C. At this stage, the fiber has a low modulus (35 GPa, 5 Msi), a low tensile strength and a low density (1.3 g/cm³).

The degree of preferred orientation can be considerably increased by stretching the carbonized fiber at very high temperature (~ 2700 - 2800°C), resulting in a high-strength and high-modulus fiber. However, this requires complicated and expensive equipment, the process is costly, and the yield of unbroken fiber is low. For these reasons, stretched fibers are no longer produced since they cannot compete with lower-cost, PAN-based fibers. As mentioned in Sec. 2.2 above, PAN is stretched prior to carbonization, which is a considerably cheaper and more reliable than stretching at very high temperature.

The low-modulus rayon-based fibers are the only ones now produced in the form of carbon cloth or felt (Thornel WCA, VCL, VCK, and VCX from Amoco Performance Products). Primary uses are in carbon-carbon composites and high-temperature insulation.

5.0   CARBON FIBERS FROM VAPOR-PHASE (CVD) REACTION

The direct growth of carbon fibers from the vapor-phase has been investigated for a number of years and the potential for producing a economically viable material with properties matching those of existing PAN or pitch-based fibers appears good.[20]-[23]

Vapor-phase fibers are produced by the catalytic decomposition of a hydrocarbon such as methane or benzene. The seed catalysts are iron particles or iron metallo-organics such as ferrocene, $(C_5H_5)_2Fe$. Growth occurs in the temperature range of 1000 - 1150°C.

The fibers still have a large spread in their tensile strength (3000 - 8000 MPa). However, the higher values compare favorably with those of high-strength PAN-based fibers (see following section).

Vapor-phase fibers are only produced in short lengths at the present time. Maximum reported length is 50 mm with diameters from 0.5 - 2 μm. Such short-length fibers would be suitable for the random reinforcement of composites and in the production of carbon-carbon (see Ch. 9).

6.0 PROPERTIES OF CARBON FIBERS

6.1 The Filament Bundle

Carbon fibers are produced as a multifilament bundle known as a *tow*. The number of filaments per tow is 500, 1,000, 3,000, 6,000 or 12,000. The smaller tow sizes are usually reserved for weaving and braiding while the larger ones are for unidirectional tape winding. Both are used primarily in aerospace applications. Still larger tows, with filament count up to 320,000, are also produced but mainly for less-demanding applications such as sporting goods (see Ch. 9).[8]

6.2 Fiber Testing

Carbon fibers are difficult to test due to their anisotropic structure, their brittleness, the variation in their diameter, and the need to mold them in an epoxy matrix to be able to measure some properties. Furthermore, the strength is dependent to some degree on the length and diameter of the test specimen and on the testing techniques. As a rule, the longer the specimen and the larger the fiber diameter, the lower the results, as shown in Fig. 8.13.[7] This is due to the greater chance of having structural defects in the larger specimens.

These factors must always be considered when comparing properties from various groups of fibers and the data shown in the following sections is to be viewed with this in mind.[24]

6.3 Physical Properties of PAN-Based Carbon Fibers

PAN-based carbon fibers are heat-treated to various degrees of structural re-ordering. This determines the final strength and modulus of elasticity. The fibers are commonly divided into the three following classes based on the value of the modulus:

- Standard-modulus fibers
- Intermediate-modulus fibers (also known as Type II)
- High-modulus fibers (also known as Type I)

Their properties are summarized in Table 8.6.

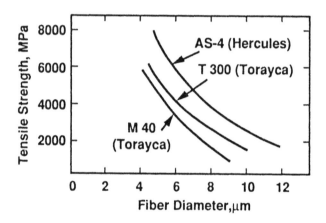

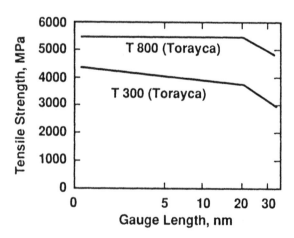

Figure 8.13. Effect of fiber diameter and gauge (specimen) length on the tensile strength of PAN-based carbon fibers.[7]

Table 8.6. Summary of Strength Properties of PAN-Based Carbon Fibers

	Standard Modulus	(Type II) Intermediate Modulus	(Type I) High Modulus
Modulus, GPA	205 - 235	275 - 310	345 - 550
Msi	30 - 34	40 - 45	50 - 80
Tensile strength, MPa	3450 - 4650	4350 - 6900	1860 - 4140
ksi	500 - 675	630 - 1000	270 - 600
Tensile strain, %	1.4 - 1.6	1.6 - 2.2	0.81 - 0.9
Density, g/cm³	1.76 - 1.79	1.76 - 1.79	1.87

Properties of selected commercially available fibers are shown in Table 8.7. The data are obtained from suppliers' technical brochures.

As a rule, higher-modulus fibers have lower tensile strength and tensile strain (elongation). The compressive-failure strain is dependent on the modulus; it increases with decreasing modulus.[25] The failure occurs by kinking or microbuckling. This tendency is shown in both PAN- and pitch-based fibers. The mean-failure strain for PAN-based fibers is 2.11 - 2.66%. It is much lower for pitch-based fibers: 0.24 - 0.98%; this may be the result of the greater suceptibility of these fibers to defects and handling. The testing procedure is described in Ref. 25.

6.4   Physical Properties of Pitch-Based Carbon Fibers

Carbon fibers based on isotropic pitch have low strength and modulus with tensile strength averaging 870 - 970 MPa and modulus of 40 - 55 GPa (Data from Carboflex, Ashland Oil Co., Ashland, KY, and Kureha Chemicals, Japan).

Mesophase pitch-based carbon fibers generally have the highest stiffness of all carbon fibers with modulus of elasticity up to 965 GPa (140 Msi), considerably higher than PAN-based fibers. The tensile strength however is much lower, averaging only half. As mentioned in the previous section, the compressive-failure strain is low.

Table 8.7. Strength and Modulus of Commercial Carbon Fibers (Partial Listing)

Product	Tensile strength Mpa	ksi	Modulus GPa	Msi
Standard Modulus				
AS-4 (1)	3930	570	248	36
Celion G30-500 (2)				
Thornel T-300 (3)	3650	530	230	33
Torayca T-300 (4)	3525	512	230	33
Grafil 33-650 (5)	4480	650		
Intermediate Modulus (Type II)				
IM-6 (1)	4340	630	275	40
IM-7 (1)	5030	730	275	40
IM-8 (1)	5860	850	275	40
IM-9 (1)	6890	1000	275	40
Celion G40 - 600 (2)	4140	600		
Celion G40 - 700 (2)	4820	700		
Celion G40 - 800 (2)	5510	800		
Thornel T-650/42 (3)	4820	700	298	42
Hitex 46-8 (6)	5760	825	296	43
High Modulus (Type I)				
UHM (1)	4140	600	448	65
Celion GY-70 (2)	1860	270	520	75
Torayca M60 (4)	2410	350	550	80

(1) Product of Hercules, Magma, UT
(2) Product of BASF, Germany, and Charlotte, NC
(3) Product of Amoco Performance Products, Atlanta, GA
(4) Product of Toray, Japan
(5) Product of Grafil, Sacramento, CA
(6) Product of BP Chemicals, UK, and Santa Ana, CA

Table 8.8 lists the range of property values of some commercially available fibers (P-series from Amoco Performance Products, Atlanta, GA).

The strength and modulus of pitch-based carbon fibers increase with increasing heat-treatment temperatures as shown in Fig. 8.14 .[11]

Table 8.8. Summary of Physical Properties of Mesophase Pitch-Based Carbon Fibers

Tensile modulus, GPa	380 - 827
Msi	55 - 120
Tensile strength, MPa	1900 - 2370
ksi	274 - 350
Tensile strain, %	0.25 - 0.5
Density, g/cm^3	2.0 - 2.18

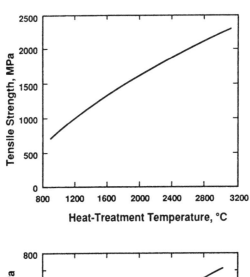

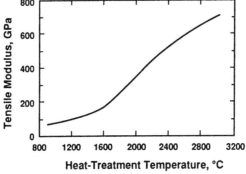

Figure 8.14. Tensile strength and modulus of pitch-based carbon fibers as a function of temperature.[11]

6.5   Properties of Rayon-Based Carbon Fibers

The properties of stretched-graphitized rayon-based carbon fibers are shown in Table 8.9. The data is to be considered for its historical value, since the material is no longer produced commercially (data from Union Carbide Corp.)

Table 8.9. Summary of Physical Properties of Rayon-Based Carbon Fibers

Tensile modulus,	GPa	173 - 520
	Msi	25 - 75
Tensile strength,	MPa	1200 - 2650
	ksi	180 - 385
Density, g/cm^3		1.40 - 1.80

6.6   Thermal and Electrical Properties of Carbon Fibers

Thermal Conductivity. As mentioned above, the fibers with the highest degree of orientation such as the pitch-based fibers have the highest thermal conductivity. As shown in Table 8.10, their conductivity along the axis is higher than even the best metal conductor. PAN-based fibers, on the other hand, have much lower conductivity because of their more pronounced isotropic structure.

Thermal Expansion. The thermal expansion of carbon fibers, measured along the axis, is extremely low and similar to that of pyrolytic graphite in the *ab* direction, i.e., slightly negative at room temperature and slowly increasing with increasing temperature (see Ch. 7, Fig. 7.11). The thermal coefficient of expansion (CTE) at room temperature is as follows (data from Amoco Performance Products).

PAN-based fibers: -0.6 to -1.1 m/m·K x 10^{-6}

Pitch-based fibers: -1.3 to -1.45 m/m·K x 10^{-6}

Table 8.10. Thermal Conductivity of Carbon Fibers and Selected Metals

Material	Thermal Conductivity at 25°C (W/m·K)
Silver	420
Copper	385
Pyrolytic graphite (ab directions)	390
PAN-based fibers* (along the axis)	8 - 70
Pitch-based fibers* (along the axis)	530 - 1100

*Data from Amoco Performance Products

Electrical Resistivity. Like the thermal properties, the electrical resistivity of carbon fibers, measured along the axis, is similar to that of pyrolytic graphite in the *ab* direction and approximately an order of magnitude higher than metal conductors such as aluminum or copper, as shown in Table 8.11.

Table 8.11. Electrical Resistivity of Carbon Fibers and Selected Metals

Material	Electrical Resistivity at 25°C ($\mu\Omega$·m)
Aluminum	0.026
Copper	0.017
Pyrolytic graphite ab directions	2.5 - 5
Pitch-based fibers* along the axis	2.2 - 2.5
PAN-based fibers* along the axis	9.5 - 18

*Data from Amoco Performance Products

REFERENCES

1. Donnet, J-B. and Bansal, R. C., *Carbon Fibers*, Marcel Dekker Inc., New York (1984)

2. Honda, H., *Carbon*, 26(2):139-136 (1988)

3. Lewis, I. C., and Lewis, R. T., *Carbon*, 26(5):757-758 (1988)

4. *Data Bank,* Gorham Advanced Material Institute, Gorham, ME (1992)

5. Fitzer, E. and Heine, M., in *Fibre Reinforcements for Composite Materials,* (A. R. Bunsell, ed.), Elsevier (1988)

6. Reisch, M. S., *C&EN*, 9-14 (Feb.2, 1987)

7. Fitzer, E., *Carbon*, 27(5):621-645 (1989)

8. Stevens, T., *Materials Engineering*, 35-38, (Aug. 1990)

9. Gupta, P. K., in *Fibre Reinforcements for Composite Materials,* (A. R. Bunsell, ed.), Elsevier (1988)

10. Jiang, H., et al., *Carbon*, 29(4&5):6353-644 (1991)

11. Riggs, D. M., Shuford, R. J., and Lewis, R. W., in *Handbook of Composites,* (G. Lubin, ed.), Van Nostrand Reinhold Co., New York (1982)

12. Edie, D. D., Fox, N. K., and Barnett, B. C., *Carbon*, 24(4):477-482 (1986)

13. Bhat, G. S., et al., *Carbon*, 28(2&3):377-385 (1990)

14. Ergun, S., in Vol. 3, Ruland, W., in Vol. 4, McKee, D. W., and Mimeault, V. J., in Vol. 8, Bacon, R. in Vol. 9, Reynold, W. N., in Vol. 11, *Chemistry and Physics of Carbon,* (P. L. Walker, Jr. and P. Thrower, eds.), Marcel Dekker, New York (1973)

15. Oberlin, A. and Guigon, M., in *Fibre Reinforcements for Composite Materials,* (A. R. Bunsell, ed.), Elsevier (1988)

16. Schulz, D. A., *SAMPE Journal*, 27-31 (Mar/Apr 1987)

17. Mochida, I., et al., *Carbon*, 28(2&3):311-319 (1990)

18. Hamada, T., et al., *J. Mater. Res.*, 5(3):570-577 (Mar. 1990)

19. Mochida, I., et al., *Carbon*, 28(1):193-198 (1990)

20. Tibbetts, G. G., *Carbon*, 27(5):745-747 (1989)

21. Masuda, T., Mukai, S. R., and Hashimoto, K, *Carbon*, 30(1):124-126 (1992)

22. Benissad, F., et al., *Carbon*, 26(1):61-69 (1988)

23. Sacco, A., Jr., *Carbon Fibers Filaments and Composites*, (J. L. Figueiredo, et al., eds.), 459-505, Kluwer Academic Publishers, Netherlands (1990)

24. Hughes, J. D., *Carbon*, 24(5):551-556 (1086)

25. Prandy, J. M. and Hahn, H. T., *SAMPE Quarterly*, 47-52 (Jan. 1991)

9

Applications of Carbon Fibers

1.0 CARBON-FIBER COMPOSITES

In the previous chapter, the processing and properties of carbon fibers were examined. The present chapter is a review of the applications and market for these fibers.

1.1 Structural Composites

A sizeable proportion of the applications of carbon fibers is found in structural composites. These composites comprise a network of fibers providing strength and stiffness and a matrix holding the fiber network together. In the so-called "advanced" or "high-performance" composites, the fibers are silicon carbide, mullite, boron, alumina, and, of course, carbon. These fibers are all competing with each other for a portion of the structural-composite business but, with increasing frequency, carbon fibers are preferred because of their low density, high strength, high modulus, and decreasing cost (see Ch. 8, Sec. 1).

Carbon fibers however are not the universal panacea, and they have several drawbacks which makes them unsuitable for many applications: they are brittle and have low impact resistance and, as a result, are difficult to weave. They also have a coefficient of thermal expansion smaller than most matrix materials, and this mismatch may cause internal stresses in the composite. In addition, they oxidize readily and are not suitable for operation at high temperature in an oxidizing atmosphere. The matrix of carbon-fiber composites can be a polymer (resin), a ceramic, a metal, or carbon itself (carbon-carbon). These matrix materials are described in Secs. 3.0, 4.0 and 5.0 below.

1.2 The Carbon-Fiber Composite Industry

The development of carbon-fiber composites has been rapid in the last twenty years and the industry is now of considerable size and diversity.[1][2] In 1991, the worldwide market for these composites was estimated at approximately $700 million, divided into the following sectors, each shown with its approximate share of the business:

Aerospace	70 %
Sporting goods	18 %
Industrial equipment	7 %
Marine	2 %
Miscellaneous	3 %

Miscellaneous applications include automotive, civil structures, mass transportation, medical products, and other consumer products.

Providing that the cost can be further reduced, the share of non-aerospace segments should increase, especially in the automotive industry.

1.3 Carbon-Fiber Composites in Aerospace

Carbon-fiber composites are found in many new structural applications such as racing cars, fishing poles, tennis rackets, competition skis, and sailboat spars. However, their greatest impact is in the aerospace industry with applications in the space shuttle, advanced passenger airplanes, aircraft brakes, and many others.

The extensive use of carbon fiber composites in aerospace is illustrated in Fig. 9.1. This figure shows the large number of applications of polymer/carbon-fiber composites in a new passenger plane, the McDonnell-Douglas Aircraft MD-12X. Other new airplanes, such as the Boeing 777 and the Airbus A340, make similar extensive use of these composites.[3] The Airbus A340, for instance, incorporates 4000 kg of epoxy-carbon fiber structures, including both vertical and horizontal stabilizers. In addition, these new airplanes have carbon-carbon brakes with considerable weight saving over conventional brakes.

The introduction of carbon-fiber composites has been slower in other areas such as the automotive industry where cost is a major factor and weight is not as critical as it is in aerospace applications.

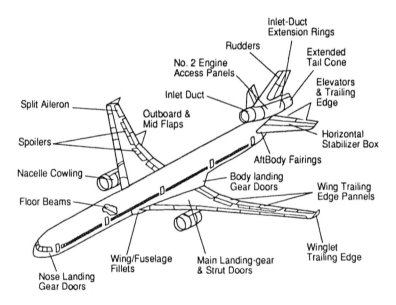

Figure 9.1. Carbon-fiber/epoxy composites in the MD-12X airplane.[3]

2.0   CARBON-FIBER ARCHITECTURE

2.1   General Characteristics

The arrangement of carbon fibers within a composite should be such that the fibers bear the loads most efficiently, usually in more than one direction. This selective reinforcement can also be enhanced by using two or more types of carbon fiber. For instance, a high-strength type might be selected to bear loads in one direction while a high-modulus type may be placed for high stiffness in another direction. An ample selection of carbon-fiber architecture is now available as a result of recent advances in sizing and weaving technology. However, a carbon fiber is inherently brittle and cannot be bent over a small radius without breaking. Consequently, the use of complicated weaving procedures such as knitting and braiding is limited.

Carbon-fiber architecture can be divided into four categories: discrete, linear (continuous), laminar (two-dimensional weave), and integrated (three-dimensional weave). The characteristics of each category are shown in Table 9.1.[4]

Table 9.1. Carbon-Fiber Architecture

Type of Reinforcement	Textile Construction	Fiber Length	Fiber Orientation	Fiber Weave
Discrete	Chopped fibers	Short	Random	None
Linear	Filament yarn	Continuous	Linear	None
Laminar	Simple fabric	Continuous	Planar	2D
Integrated	Advanced fabric	Continuous	3D	3D

2.2 Yarn and Roving

A carbon-fiber yarn is an assembly of monofilaments held together by a twist. Yarns are usually composed of continuous filaments or, in some cases, of discrete filaments (staple yarns). Woven fabrics are usually processed from yarns comprising several thousand monofilaments.

A carbon fiber roving is a continuous fiber bundle with essentially no twist, usually containing more monofilaments than a yarn.

2.3 Discrete Fibers

Discrete fibers (also known as chopped fibers) are short-length fibers (a few centimeters) which are generally randomly oriented. They are usually low-strength and low-cost fibers in the form of felt or mat, with applications in special types of carbon-carbon and in high-temperature insulation (see Secs. 4.0 and 7.1 below).

2.4 Continuous Filaments

Most carbon fibers are in the form of continuous filaments with a diameter averaging 10 μm. They are applied unidirectionally (0°) by the processing techniques of filament winding and tape layup described in Ref. 2. Such unidirectional systems have the highest property-translation efficiency, i.e., the fraction of fiber properties translated into the composite. On the other hand, they have low interlaminate strength because of the lack of fibers in the thickness direction.

2.5 Laminar (2D Weaves)

Woven carbon fibers are usually biaxial structures, woven at 0° and 90° (warp and fill) in three basic patterns: plain, satin, and twill. The highest frequency of yarn interlacing is found in the plain weave, followed by the twill and the satin weave. In the satin weave, the warp ends are woven over four fill yarns and under one (five-harness satin) or over seven and under one (eight-harness satin). The property translation efficiency is the highest in satin weave, followed by twill and plain weave. Fig. 9.2 shows the plain and twill weaves.[4]

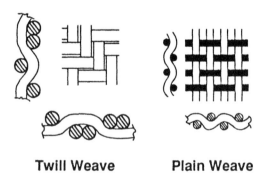

Twill Weave   Plain Weave

Figure 9.2. Weaving patterns of biaxially woven fabrics.[4]

2.6 Integrated (3D Weaves)

As mentioned above, the brittleness of the carbon fiber puts a limit on three-dimensional processing. A recently developed integrated (3D) weave consists of an orthogonal non-woven fabric produced by placing fibers in three or more orthogonal directions (i.e., mutually perpendicular) with no interlacing. Strength in the thickness direction is high but the weaving equipment is complicated, cost is considerable, and shapes are limited.[5]

3.0 CARBON-FIBER POLYMER (RESIN) COMPOSITES

3.1 Polymer (Resin) Matrices

The most common matrix materials of carbon-fiber composites are the polymers, also called resins or plastics. Carbon-reinforced polymers are low-density, high-strength, and high-modulus composites with extensive applications, especially in aerospace as mentioned above. Their cost is still high but is gradually decreasing as the fabrication techniques are becoming less labor-intensive.

A number of polymers are suitable as matrix material, each with its own advantages and disadvantages, with wide differences in properties, and the selection of a given carbon fiber-polymer system must be made after a thorough analysis of its suitability for the application.

The polymers are usually processed in a preliminary step in the form of "prepreg", that is pre-coated and partially cured (polymerized) on the fiber. This "prepreging" provides uniform impregnation of the fiber bundle and uniform resin-to-fiber ratio. The following polymers are presently available commercially.[2][6]

Epoxy Polymers. Epoxy polymers provide high-strength matrix but are usually limited to room-temperature applications, unless a high-temperature curing agent is used, in which case good performance can be expected up to 150°C. A drawback of carbon-fiber epoxy laminates is their low-impact resistance. This can be offset to some degree by the addition of a thermoplastic modifier to the epoxy (toughened epoxy).

High-Temperature Polymers. Several polymers with higher-temperature capability than epoxies are now available in the form of prepreg. Their maximum-use temperature is shown in Table 9.2.

Table 9.2. Maximum-Use Temperature of Polymer Matrices[2]

Polymer	Maximum-Use Temperature, °C
Epoxies	up to 150
Bismaleimides-epoxies (BMI)	205 - 245
Polyimides (PI)	260 - 315
Polybenzimidazoles (PBI)	315 - 370

Thermoplastic Polymers. Thermoplastic polymers do not require a cure cycle but need only to be melted during processing (usually injection molding). The most common are nylon, polypropylene, and polyethylene which are usually molded with 10 - 25 vol.% discrete (chopped) carbon fibers. The addition of fibers substantially increases the modulus and, to a lesser degree, the strength. Electrical conductivity is also considerably increased and many applications of these composites are found in *electromagnetic-interference (EMI)* shielding.

The major drawback of carbon-fiber thermoplastic composites is their low-temperature resistance.[7][8] However, recently developed thermoplastic polymers have much higher temperature resistance and are being considered as matrices for continuous-fiber composites. These polymers include polyethersulfone (PES), polyetheretherketone (PEEK) and polyphenyl sulfide (PPS).[2] PEEK in particular has excellent potential since it is less brittle than the epoxies and provides a tougher composite (see Table 9.5 below).

The various fabrication techniques for carbon-fiber composites include filament winding, injection and compression molding, pultrusion, and wet layup. They are described in Ref. 9.

3.2 Surface Treatment of Carbon Fibers

The surface of a carbon fiber (or of diamond, graphite, and any other crystalline solid) has been described as an extreme case of lattice defect.[10] The regular configuration of carbon atoms ends abruptly and the surface atoms have a different coordination with dangling bonds which are able to react with any atom or molecule present on the surface. The result is the formation of compounds such as basic or acidic surface oxides, CO_2, and others, as shown in Fig. 9.3.[11] The greater the degree of graphitization of the fiber, the less surface reaction there will be since the surface area of a heat-treated fiber is many times smaller than that of the untreated fiber.

It is possible to take advantage of these surface properties to improve the adhesion between the carbon fiber and the polymer matrix by treating the surface and coating it with a coupling agent. Such a surface treatment is generally an oxidation process which can be wet, dry, or anodic. Coupling agents include copolymers of maleic anhydride, pyrolytic graphite, or polyimide.[11]-[13]

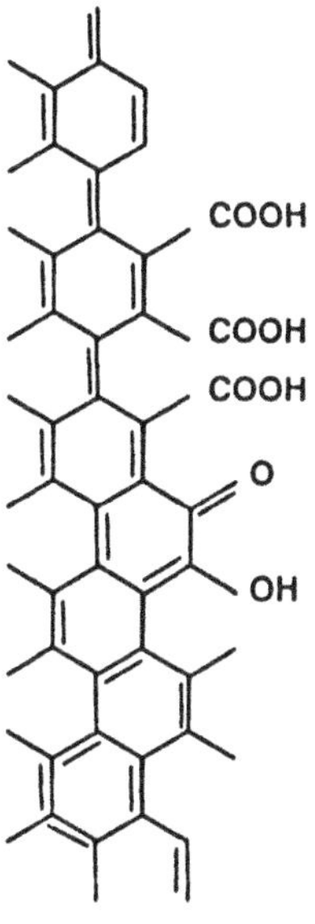

Figure 9.3. Types of solid surface oxides on carbon fibers.[12]

3.3   Properties of Carbon-Fiber Polymer Composites

As seen in the previous sections, a carbon-fiber composite has a complex nature and its properties may vary widely as a function of the type of polymer matrix, the processing variables (i.e., temperature, post-cure, etc.), the ratio of fiber to matrix, the type and properties of the carbon fiber, the orientation of the fibers, and the fiber architecture in general.

A comprehensive study of the effect of all these variables on the properties of the composite is a formidable task which has yet to be undertaken. Yet a good deal of information is already available, part of which is summarized in the following three tables. The data are compiled from the literature and suppliers' bulletins. Table 9.3 lists the effect of the type of carbon fiber.

Table 9.3. Effect of the Type of Carbon Fiber on the Properties of Carbon-Epoxy Composites[14]

| | Type of Fiber | | |
	T650/42*	P55**	P120***
Density, g/cm³	1.6	1.7	1.8
Tensile strength, MPa	2585	896	1206
Modulus, GPa	172	220	517
Compressive strength, MPa	1723	510	268
Shear strength, MPa	124	55	27
Thermal conductivity, W/m·K	8.65	74	398

    *   T650/42: a high-strength, intermediate-modulus, PAN-based fiber
    ** P55: an intermediate-modulus, pitch-based fiber
    *** P120: a high-modulus, pitch based fiber

Notes: 1. The fibers are produced by Amoco Performance Products Inc.

2. The composites have the same epoxy matrix with 62 vol.% fiber.

3. Fiber orientation is unidirectional (0°) and tested in the fiber direction.

Particularly noticeable is the high thermal conductivity of the graphitized pitch-based fiber composite which is 46 times that of the PAN-based material.

Table 9.4 shows the effect of fiber orientation on the properties. The large differences in tensile strength and modulus of unidirectional laminates tested at 0 and 90° should be noted.

Table 9.4. Effect of the Orientation of Carbon Fibers on the Properties of Carbon Polymer Composites[15]

Testing angle	Unidirectional Laminate		Quasi-isotropic Laminate	
	0°	90°	0°	90°
Tensile strength, MPa	793	20.0	379	241
Tensile modulus, GPa	303	3.3	104	97
Tensile ultimate strain, %	0.25	0.5	0.27	0.23
Compression strength, MPa	400	158	172	200
Compression modulus, GPa	255	6.7	76	88
Compressive ultimate strain, %	-	-	0.55	0.86

Notes:
1. The carbon fiber is Amoco P75 high-modulus pitch fiber
2. Fiber content is 60 vol.%
3. Polymer is epoxy PR500-2 from 3M
4. Isotropic laminate has 0°, 30°, 60°, 90°, 120°, 150° stacking sequence
5. Unidirectional laminate tested along the fibers (0°) and across the fibers (90°). Isotropic laminate tested in two directions perpendicular to each other

Table 9.5 shows the effect of two types of polymer matrices, epoxy and PEEK, on the mechanical properties of a carbon-fiber composite with two different fiber orientations: 0° (unidirectional) and 45° (three layers at 0°, 45° and 90° respectively). The fracture strain of the PEEK 45° composites is considerably larger than that of the others.

3.4 Applications of Carbon-Fiber Polymer Composites

The number of applications of carbon-fiber polymer composites is growing rapidly in aerospace, sporting goods, and industrial areas. Table 9.6 lists some of these typical applications which are in production or still experimental.[1][3][16]

Table 9.5. Mechanical Properties of Carbon-Fiber Composites with Epoxy and PEEK Polymer Matrices[2]

Polymer Matrix	Fiber Orientation	Tensile Strength (MPa)	Tensile Modulus (GPa)	Fracture Strain %
Epoxy	0°	932	83	1.1
Epoxy	45°	126	-	1.3
PEEK	0°	740	51	1.1
PEEK	45°	194	14	4.3

Table 9.6. Typical Applications of Carbon-Fiber Polymer Composites

- Aerospace
  Primary aircraft structure, wing, fuselage
  Control surfaces
  Vertical stabilizer (A310 Airbus, Boeing 777)
  Satellites bodies
  Space-shuttle manipulating arm
  Satellite solar-array, wings, and substrate
  Space antennas and reflectors
  Voyager-spacecraft high-gain antenna
  Optical benches for space telescopes
  Space mirrors
- Sporting Goods
  Tennis rackets
  Skis and golf clubs
  Sailboat spars
- Electrical and Electrochemical
  Electromagnetic-interference shielding
  Battery electrodes
- Biomedical
  Bone prostheses
- Musical
  Soundboard and body for acoustical guitar

4.0 CARBON-CARBON

4.1 General Characteristics of Carbon-Carbon

The carbon-fiber/polymer composites reviewed in the previous section have excellent mechanical properties but limited temperature resistance. Maximum operating temperature is presently ~ 370°C (Table 9.2). These composites cannot meet the increasingly exacting requirements of many aerospace applications which call for a material with low density, excellent thermal-shock resistance, high strength, and with temperature resistance as high or higher than that of refractory metals or ceramics. These requirements are met by the so-called "carbon-carbon" materials.

Carbon-carbon refers to a composite comprising a carbon-fiber reinforcement and a carbon matrix, in other words an all-carbon material. Carbon-carbon was developed in the early 1960's in various programs sponsored by the United States Air Force.[17][18] Some of the early applications of carbon-carbon were in nose tips and heat shields of reentry vehicles.[19][20] Carbon-carbon is now a major structural material, not only in aerospace, but also in many non-military applications.

Carbon-carbon has many common features with the various materials already reviewed in previous chapters such as molded carbon and graphite, polymeric carbon and pyrolytic graphite. It is often considered as a refractory version of the carbon-fiber/ polymer composites.

Carbon-carbon has a major disadvantage, that is poor oxidation resistance and efforts to solve the problem have met with only limited success so far. A solution will have to be found to this Achilles' heel before the material can achieve its full potential.

4.2 Carbon-Carbon Composition and Processing

Carbon-Fiber Network. Rayon-based carbon fibers were used in the early development of carbon-carbon and are still used as carbon felt. PAN-based fibers are now used extensively and pitch-based fibers are under investigation.[21] The selection of the carbon-fiber architecture is determined by the application and include felt, short (chopped) fibers, continuous filament such as small-tow T-300 fiber, filament winding or tape-layup, and 3D structures (see Sec. 2.0 above). The effect of carbon-fiber type and architecture is reviewed in Refs. 22 and 23. The effect of carbon-fiber

surface treatment on the mechanical properties of carbon-carbon composites is complex and still not clearly understood at this time.[24]

Matrix Materials and Processing. In most cases, the starter matrix
material is a high-carbon-yield polymer such as a phenolic (reviewed in Ch.
6, Sec. 2.1). Other polymers are under investigation, including
polyarylacetylene and aromatic diacetylene, the latter giving an unusually-
high char yield of 95%.[25][26] Most polymers are considered essentially non-
graphitizable (see Ch. 4, Sec. 3.5).

Processing is lengthy, difficult, and expensive. Typical steps in the
processing of a 2D phenolic-carbon fiber composite are as follows:

1. Vacuum bag/autoclave cure to $150°C$ in an 8-hour cycle

2. Rough trimming and inspection

3. Post-cure in a restraining fixture to $265°C$ for 7 days

4. Pyrolysis (carbonization) in a steel retort packed with
   calcined coke in an inert atmosphere, the cycle consisting
   of 50 hours to reach $820°C$, 20 hours at $820°C$ and 20
   hours cool-down

5. Impregnation with furfuryl alcohol in an autoclave, followed
   by a 2-hour cure at $150°C$ and a 32-hour post-cure at
   $210°C$

6. Pyrolysis, repeat of step 4

7. Impregnation, cure, and pyrolysis cycles repeated for a
   total of three times

8. Graphitization heat treatment ($>2000°C$) if required

9. Coating with silicon carbide (SiC) for oxidation protection
   (optional)

Processing may vary depending on the characteristics of the starter
materials, the geometry of the part, and other variables.[27] Pyrolyzation
temperature often reaches $1000°C$. The chemistry of the pyrolysis process
is reviewed in Ch. 4, Sec. 2.0.

Chemical Vapor Infiltration: Instead of a carbonized polymer, the
matrix material can be a pyrolitic carbon obtained by the thermal decomposition of a hydrocarbon gas, usually methane (CH_4) or propane (C_3H_8). This
process is known as chemical vapor infiltration (CVI) and is described in Ch.
7, Sec. 2.6. Various CVI techniques such as temperature and/or pressure

gradients can be selected depending on the application.[23][30]-[32] CVI is used in the fabrication of aircraft brakes (see below).[29]

Impregnation (Densification). The material experiences considerable shrinkage during pyrolysis due to the elimination of non-carbon components. This causes cracking and a high-degree of porosity. To optimize the properties, a densification step is necessary. It is accomplished by liquid impregnation or by CVI.

Liquid-impregnation materials such as furfuryl-alcohol polymer or low-viscosity pitch are forced into the pores by the application of a pressure gradient and then cured. A proper densification may require numerous impregnation/cure cycles and inordinately long (many hundred hours) processing. However, the processing time can be shortened considerably by heating the preform inductively while immersed in the liquid hydrocarbon. This leads to rapid infiltration and carbon deposition within the structure rather than on the outside surfaces.[28]

Densification can also be achieved by CVI. As a general rule, polymer/pitch liquid infiltration is more effective and usually less costly for thick parts, while CVI provides higher strength and modulus and is best suited for thinwall parts.[31][32]

4.3   Properties of Carbon-Carbon Composites

As mentioned in Sec. 3.3 above, a carbon-fiber composite has a complex nature with wide differences in properties. This is certainly true of carbon-carbon.[33] The task of analyzing the fabrication variables and their effects on the properties of the composite is far from completed.

Some of the available data, compiled from the literature and suppliers' bulletins, are summarized in Table 9.7. The referenced materials are: *(a)* a 2D layup of woven cloth made from PAN-based carbon fibers (ACC from the Vought Corp., Dallas, TX), and *(b)* a square-weave carbon fabric layup, also from PAN-based carbon fibers (BP Chemicals, Gardena, CA).[34][35]

These values show that the strength and modulus that could be expected in the composite as a result of the high strength and high modulus of the carbon-fiber reinforcement have yet to be realized. In fact, they are only 20 to 50% of the rule-of-mixtures predictions. Progress is being slowly made in improving these values but carbon-carbon is still far from achieving the 90 - 95% rule-of-mixtures predicted strength of some carbon-fiber/epoxy composites.[36]

Table 9.7. Selected Properties of Carbon-Carbon Composites

	ACC	Semicarb
Density, g/cm^3	1.84	1.7
Coef. of Thermal Expansion (0-1000°C), (x 10^{-6})/°C	0.3	0.7
Thermal Conductivity at 25°C, W/m·K	6.9	33.8
Tensile Strength: MPa	165	270
ksi	24	40
Tensile Modulus: GPa		110
Msi		16
Flexural Strength: MPa		303
ksi		44

Yet, the strength of carbon-carbon is considerably higher than that of the strongest grades of molded graphites. For instance, a high-density isostatically-pressed graphite has a flexural strength of 77 MPa (Ch. 5, Table 5.4) compared to 303 MPa for carbon-carbon. However, these graphites have a generally higher thermal conductivity (159 W/m·K for petroleum-coke electrographite) and a higher CTE (3.2 - 7.0 [x 10^{-6}]/°C) than the carbon-carbon composites (Ch. 5, Tables 5.7 and 5.8).

A major advantage of carbon-carbon composites over other structural materials is their high-temperature mechanical properties which, like other graphitic materials, does not decrease with increasing temperature. The specific strengths of carbon-carbon and other materials as a function of temperature are shown in Fig. 9.4.[37]

Oxidation Protection. Like all carbon and graphite materials, carbon-carbon is susceptible to oxidation and oxidizes rapidly above 500°C. The principal oxidation protection material is silicon carbide (SiC) which is usually applied by *pack cementation.* Molten silicon reacts with the carbon surface to form SiC and surface pores are filled to a depth of approximately 0.5 mm. SiC has higher thermal expansion than carbon-carbon and the coating has a tendency to crack during cool-down. Although partial protection is provided, the oxidation problem has yet to be solved, particularly at high temperatures.[18][37]

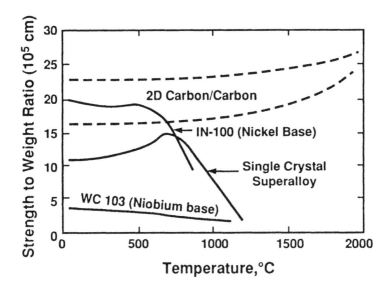

Figure 9.4. Specific strength of carbon-carbon and refractory alloys as a function of temperature.[37]

4.4 Carbon-Carbon Applications

Table 9.8 lists some of the present applications of carbon-carbon. These are either in production or in the development stage.[28][38]

Aircraft Brakes. By far the largest volume application of carbon-carbon is found in aircraft brakes. Carbon-carbon brakes offer a lifelong, lightweight alternative to steel brakes. For instance, the weight saving on the Boeing 767 is 395 kg with double the number of landings per overhaul. Since the coefficient of friction remains constant with temperature, brakes do not fade. A typical fabrication process includes carbonizing PAN-fiber fabrics to 1000°C, cutting the fabric to shape, impregnating with a polymer, carbonizing, and densifying by CVD by the decomposition of natural gas at low-pressure.[29] A carbon-carbon brake assembly is shown in Fig. 9.5.

Cost of Carbon-Carbon. Because of the long and involved processing, the cost of carbon-carbon is high and estimated at over $200/kg for industrial parts and considerably more for some aerospace applications.

Table 9.8. Applications of Carbon-Carbon

- Structural parts for hypersonic vehicles
- Nose tips and heat shields of reentry vehicles
- Space-shuttle leading edges, nose cap (these parts are coated with SiC for oxidation protection)
- Conversion flaps, combustion liners
- Rocket-nozzle throat sections and exit cones
- Pistons and connecting rods for internal-combustion engines (experimental)
- Heating elements, with 2000°C working temperature
- Brakes for aircraft and spacecraft
- Racing-car brakes and clutch plates
- Hardware for low-pressure and plasma CVD such as cage boats, ladders, trays, liners, cage supports, baffles
- Electrode connecting parts in electrical-arc furnaces
- Metal-forming molds and blowing forms, particularly for titanium fabrication
- Interceptor for glass bottle molding (gob)

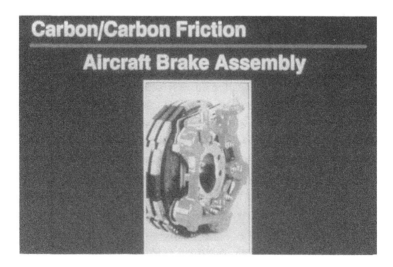

Figure 9.5. Carbon-carbon aircraft brake. *(Photograph courtesy of BP Chemicals, Advanced Materials Div., Gardena, CA.)*

5.0 METAL-MATRIX, CARBON-FIBER COMPOSITES

Carbon-fibers with metal matrices (typically aluminum or copper) are promising composites with definite advantages over carbon-fiber/polymer-matrix composites, such as high thermal and electrical conductivities and strength and dimensional stability at high temperature. In addition, their coefficient of thermal expansion (CTE) is low and can be readily tailored. However, a number of problems remain to be solved before the capability of these composites can be fully demonstrated. Yet progress is being made and some applications are reaching the commercial stage.

5.1 Fiber-Matrix Interaction

The carbon-fiber, metal-matrix interface is a critical factor. Aluminum and other metals in the molten stage tend to react with the fiber and form carbides and intermetallics. Although the resulting diffusion layer creates a useful chemical bond, it also degrades the strength of the fiber and considerably reduces its ability to reinforce the composite. The interface chemistry can be complicated and involves the formation of preferential grain-boundary attack by solid solutions, volume misfits, recrystallization, the formation of microcracks, and other detrimental phenomena. The *galvanic couple* between the metal matrix and the carbon fiber may cause considerable *galvanic corrosion* as is the case with aluminum. Copper, on the other hand, has little or no galvanic corrosion.

The diffusion problems can be partially overcome by applying a barrier coating such as titanium, boron, nickel, copper, or niobium carbide over the fiber prior to processing the composite. The latter material, NbC, has shown excellent diffusion-barrier characteristics in a laminate composed of P-100 fibers (43 vol.%) and a copper matrix.[39] A cross-section of the composite after isothermal exposure at 815°C for 240 hours is shown in Fig. 9.6.

5.2 Fabrication Process

Fabrication techniques include diffusion coating and electroplating, followed by hot-pressing and liquid-metal infiltration. In some cases, the carbon fibers are coated by electroplating copper. A carbon tow is used instead of a twisted bundle to facilitate infiltration. This is followed by plasma-spraying of additional copper and hot-pressing.[40]

A CO_2 laser-roller densification process is used to fabricate a unidirectional carbon-fiber–aluminum-matrix composite.[41] Although good densification is obtained, the mechanical properties of the composites are low, probably due to fiber degradation.

Aluminum and copper are the most common matrix materials investigated so far.[9] Other metals include nickel, niobium, and beryllium.[40][42]

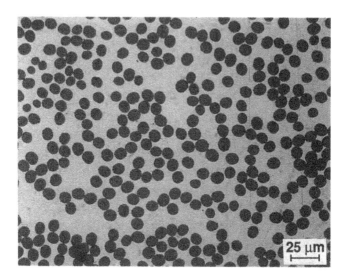

Figure 9.6. Cross-section of graphite-fiber/copper composites with NbC interface after thermal exposure. *(Photograph courtesy of Rocketdyne, Canoga Park, CA.)[39]*

5.3 Properties

Information on the properties of carbon-fiber, metal-matrix composites is still scanty with little or no suitable comparative data available. In many reports, important variables such as fiber-matrix ratio and fiber orientation are not mentioned. As a rule, the mechanical properties of present composites are still far short of the potential predicted by the rule-of-mixtures.

The thermal conductivity can be increased by the proper utilization of high-conductivity fibers such as Pitch-130 or CVD fibers (see Ch. 8, Secs. 3.0 and 5.0). The thermal conductivity of these fibers is considerably higher than that of copper and other metals as shown in Fig. 9.7.[43]

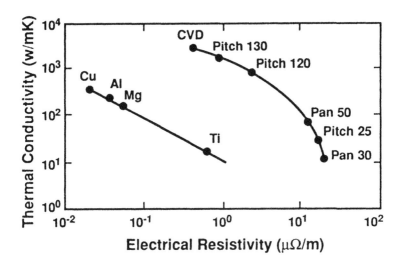

Figure 9.7. Room-temperature thermal conductivity and electrical resistivity of carbon fibers and selected metals.[43]

5.4 Applications

Potential applications of carbon-fiber metal-matrix composites are found where high thermal conductivity and increased stiffness are required. Typical applications now under consideration are listed below.[43][44]

- Heat-radiating fins in space nuclear systems
- Rocket-thrust chambers in main engine of space shuttle
- Heat-exchanger component for hypersonic vehicle shown in Fig. 9.8[39]
- High-stiffness, low thermal expansion, heat-sink plates for high-density electronic packaging

Figure 9.8. Actively cooled leading-edge panel concept for hypersonic vehicle. Material is a carbon fiber/copper composite.[39]

6.0 CERAMIC-MATRIX, CARBON-FIBER COMPOSITES

The major drawback of ceramics is their intrinsic brittleness. For example, most metals have a fracture toughness forty times greater than conventional ceramics and glasses. This brittleness is related on the atomic level to the strong hybrid-ionic-covalent bonds of ceramics. These strong bonds prevent deformation such as occurs in ductile metals. Applied stresses tend to concentrate at the sites of flaws, at voids and chemical impurities, and at grain interfaces. The result is catastrophic brittle failure.

Ceramics, reinforced with carbon fibers or whiskers, are less brittle and have increased fracture toughness and improved thermal-shock resistance. These composites have excellent potential but, as with metal-matrix composites, many problems must be solved before they become reliable engineering materials.

6.1 Matrix Materials and Fiber-Matrix Interaction

Oxides such as alumina (Al_2O_3) are generally not suitable matrix materials in carbon-fiber composites as the carbon reduces the oxide to form metal carbide and CO during the fabrication process. The oxidation of the carbon fiber may be sufficient to generate a high partial pressure of CO which results in the formation of gas bubbles and cracks in the oxide.[45]

As opposed to oxides, carbides such as silicon carbide (SiC) and boron carbide (B_4C) are compatible with carbon fibers, and satisfactory composites are produced with these matrices and PAN-based yarn by chemical vapor infiltration (CVI).[46] A boron-carbon intermediate coating provides optimum strength and toughness as it prevents fiber degradation.

The tensile strength data are shown in Table 9.9.

The carbon-fiber composites show substantially higher strength than the silicon carbide materials but degrade more rapidly in air at $1100°C$. The fracture modes of a carbon fiber/silicon-carbide matrix composite are shown in Fig. 9.9.

Table 9.9. Tensile Strength of Carbon Fiber, SiC- and B_4C-Matrix Composites

Composite	Tensile Strength MPa (ksi)	Standard Deviation MPa (ksi)
Carbon fiber, SiC matrix	515 (74.7)	58 (8.4)
Carbon fiber, B_4C matrix	380 (55)	26 (3.8)
SiC fiber,SiC matrix	310 (45)	62 (9)
SiC fiber, B_4C matrix	314 (45.6)	68 (9.8)

Notes:

1. Carbon fibers are T-300 (Amoco Performance Products).
2. Silicon carbide fibers are Nicalon (Nippon Carbon Co.).
3. Both fibers are coated with B-C intermediate.
4. Fibers are unidirectional and composite was tested in fiber direction.

Figure 9.9. Fracture of a carbon fiber/silicon-carbide matrix composite. *(Photograph courtesy Rockwell International, Canoga Park, CA.)*

6.2 Applications

Applications of carbon-fiber, ceramic-matrix composites are still essentially in the development stage and many fabrication problems must be solved before the full potential of these materials is realized. Among the more successful applications to date are the following.

Carbon-Fiber, Cement Composites. Work carried out in Japan has shown the potential of carbon fibers as an effective reinforcement for cement.[47] In one composite application, the fibers were pitch-based, either chopped, continuous, or in a mat form, and comprised 2 - 4% by volume of the composite. As shown in Fig. 9.10, the fibers impart a marked increase in tensile strength and pronounced change in fracture mode. No deterioration of the fiber occurs due to the high resistance of the carbon fiber to alkali solutions. When cured in an autoclave (as opposed to air cure), tensile strength more than triples to 6.8 MPa with 4% carbon. These composites are now produced commercially on a modest scale in Japan with applications in wall- and floor-panels, foot bridges, and other architectural components.[48]

Niobium Nitride (NbN) Superconductor: Thin films of NbN are applied on carbon fibers by vapor deposition followed by thermal-sprayed oxygen-free high-conductivity copper (OFHC). A transition temperature of the composite of 16 K and a critical current of 10^7 A/cm^2 have been observed. Applications are in high-energy lasers, particle-beam weapons, and electromagnetic guns.[49]

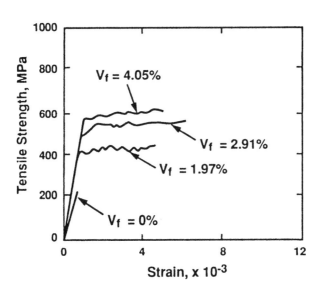

Figure 9.10. Tensile stress-strain curves of carbon-reinforced cement with various volume-fractions of chopped carbon fibers.[47]

7.0 OTHER APPLICATIONS OF CARBON FIBERS

7.1 High-Temperature Thermal Insulation

Carbon fibers in felted or woven form are the preferred materials for high-temperature insulation for furnaces operating above 1500°C in vacuum or inert atmosphere. However, these materials have a tendency to sag and compact with time which eventually results in non-uniform insulation characteristics. This problem is partially solved by the use of rigid insulation which consists of a felt impregnated with a polymer, subsequently pyrolyzed, and fired and outgassed at 2000°C under vacuum. The material is available in the form of cylinders, boards, disks and special shapes to suit furnace demands.[50]

7.2 Electrical Applications

Carbon fibers are being considered for *electromagnetic interference* (EMI) shields. The fibers are sometimes coated with nickel by electroplating to increase the electrical conductivity, and molded in the form of sheets with a polymer such as polycarbonate.[42]

7.3 Electrochemical Applications

Many potential applications of carbon fibers in batteries have been reported but their inherent high cost compared to conventional electrode materials such as graphite powder, porous carbon, and molded carbon have precluded their use in commercial applications so far (see Ch. 10, Sec. 3.0).[2]

REFERENCES

1. *Carbon Fibers, Filaments and Composites* (J. L. Figueiredo, et al., eds.), Kluwer Academic Publishers, The Netherlands (1989)

2. Dresselhaus, M. S., Dresselhaus, G., Sugihara, K., Spain, I. L., and Goldberg, H. A., *Graphite Fibers and Filaments*, Springer Verlag, Berlin (1988)

3. Stover, D., *Advanced Composites*, 30-40 (Sept./Oct. 1991)

4. Ko, F. K., *Ceramic Bulletin*, 68(2):401-414 (1989)

5. Taylor, A. H., Davis, T. S., Ransone, P. O., and Drews, M. J., *Proc. of the 14th. NASA/DoD Conf.*, NASA Conf. Publ. 3097, 2:455 (1990)

6. Recker, H. G., et al., *SAMPE Quaterly*, 46-51 (Oct. 1989)

7. Gerteisen, S. R., *Proc. 43rd SPE Annual Conf.*, 1149-1151 (1985)

8. *Electrafil Electrically Conductive Thermoplastics*, Bulletin of Akzo Engineering Plastics Inc., Evansville, IN (1989)

9. Donnet, J. B. and Bansal, R. C., *Carbon Fibers*, Marcel Dekker Inc., New York (1984)

10. Boehm, H. P., *Advances in Catalysis*, (D. D. Eley, et al., eds.), Vol. 16, Academic Press, New York (1966)

11. Fitzer, E. and Heine, M., in *Fibre Reinforcements for Composite Materials* (A. R. Bunsell, ed.), Elsevier (1988)

12. Donnet, J. B. and Guilpain, G., *Carbon*, 27(5):749-757 (1989)

13. Whang, W. T. and Liu, W. L., *SAMPE Quarterly*, 3-9, (Oct. 1990)

14. *Thornel Carbon Fibers and Composites*, Bulletin of Amoco Performance Products, Atlanta, GA (1991)

15. Blair, C. and Zakrzcwski, J., *SAMPE Quarterly*, 3-7 (April 1991)

16. Stevens, T., *Materials Engineering*, 35-38 (Aug. 1990)

17. Buckley, J. D., *Ceramic Bulletin*, 67(2):364-368 (1988)

18. Fitzer, E., *Carbon*, 25(2):163-190 (1987)

19. Pierson, H. O., *Advanced Techniques for Material Investigation*, SAMPE 14 II-4B-2 (1968)

20. Granoff, B., Pierson, H. O., and Schuster, D. M., *Carbon*, 11:177-187 (1973)

21. Manocha, L. M., Bahl, O. P., and Singh, Y. K., *Carbon*, 29(3):351-360 (1991)

22. Manocha, L. M. and Bahl, O. P., *Carbon*, 26(1):13-21 (1988)

23. Oh, S. M. and Lee, J. Y., *Carbon*, 26(6):769-776 (1988)

24. Manocha, L. M., Yasuda, E., Tanabe, Y., and Kimura, S., *Carbon*, 26(3):333-337 (1988)

25. Zaldivar, R. J., Kobayashi, R. W., and Rellick, G. S., *Carbon*, 29(8):1145-1153 (1991)

26. Economy, J., Yung, H., and Gogeva, T., *Carbon*, 30(1):81-85 (1992)

27. Weisshaus H., Kenig, S., Kastner, E., and Siegman, A., *Carbon*, 28(1):125-135 (1990)

28. Marshall, D., *Advanced Composites Engineering*, 14-16 (June 1991)

29. Awasthi, S. and Wood, J. L., *Ceram. Eng. Sci. Proc.*, 9(7-8):553-560 (1988)

30. Pierson, H. O., *Handbook of Chemical Vapor Deposition*, Noyes Publications, Park Ridge, NJ (1992)

31. Lackey, W. J., *Ceram. Eng. Sci. Proc.*, 10(7-8):577-584 (1989)

32. Starr, T. L., *Ceram. Eng. Sci. Proc.*, 9(7-8):803-812 (1988)

33. Stover, E. R. and Price, R. J., *Proc. of the 14th. NASA/DoD Conf.*, NASA Conf. Publ. 3097, 2:345-358 (1990)

34. Taylor, A. and Ransone, P., *Proc. of the 13th. NASA/DoD Conf.*, NASA Conf. Publ. 3054, 1:451-455 (1989)

35. *Advanced Reinforced Composite Materials, Semicarb*, Bulletin from BP Chemicals (Hitco) Inc., Gardena, CA (1992)

36. Binegar, G. A., et al., *Proc. of the 13th. NASA/DoD Conf.*, NASA Conf. Publ. 3054, 1:281 (1989)

37. Strife, J. R. and Sheehan, J. E., *Ceramic Bulletin*, 67(2):369-374 (1988)

38. Lewis, S. F., *Materials Engineering*, 27-31 (Jan. 1989)

39. Ash, B. A. and Bourdeau, R., *Proc. TMS Annual Meeting, High Performance Copper Materials Session*, New Orleans, LA (Feb. 1991)

40. Upadhya, K., *Journal of Metals*, 15-18 (May 1992)

41. Okumura, M. et al., *SAMPE Quarterly*, 56-63 (July 1990)

42. Evans, R. E., Hall, D. E., and Luxon, B.A., *Proc. 31st. Int. SAMPE Synposium*, 177-190 (April 1986)

43. Zweben, C., *Journal of Metals*, 15-23 (July 1992)

44. Doychak, J., *Journal of Metals,* 46-51 (June 1992)

45. Luthra, K., and Park, H. D., *J. Am. Ceramic Soc.*, 75(7):1889-1898 (1992)

46. Kmetz, M. A., Laliberte, J. M., and Suib, S. L., *Ceramic Eng. and Sci. Proc., 16th. Annual Conf. on Composites and Advanced Ceramic Materials,* 743-751, Am. Ceramic Soc., (1992)

47. Inakagi, M., *Carbon*, 29(3):287-295 (1991)

48. Plellisch, R., *Advanced Composites*, 37-41 (Jan/Feb. 1992)

49. Hunt, M., *Materials Engineering*, 25-28 (Oct. 1990)

50. Hamling, P. D., *Ceramic Bulletin*, 67(7):1186-1189 (1988)

10

Natural Graphite, Graphite Powders, Particles, and Compounds

1.0 NATURAL GRAPHITE

Natural graphite is a relatively abundant mineral found in many parts of the world and known to man for centuries. It was originally considered a separate substance, thought to contain lead and often confused with molybdenum sulfide. In 1855, Brodie prepared pure graphite and unequivocally identified it as an allotrope of carbon. Today, although the majority of graphite products are synthetic, natural graphite is still preferred for some applications.[1]

1.1 Characteristics and Properties

The characteristics and properties of natural graphite are essentially those of single-crystal graphite described in Ch. 3. The weak interlayer bond of the crystal allows easy slippage of the planes, giving graphite its pronounced softness which is graded lower than talc, namely <1 on the Moh scale.

Natural graphite is black and lustrous (that is, it reflects light evenly without glitter or sparkle) and, because of its softness, it marks other substances readily, leaving a typical black mark (see Ch. 3, Sec. 1.1).

Other characteristics can be summarized as follows:

- Low density (2.1 - 2.3 g/cm^3, depending on the type)
- Optically opaque even in extremely thin sections
- Chemically stable at ordinary temperature
- Unaffected by weathering (as evidenced by bright flakes found in disintegrated graphite-bearing rocks)
- High thermal and electrical conductivity
- Low coefficient of thermal expansion
- Low coefficient of friction
- Greasy feel

1.2 Types of Natural Graphite

Natural graphite is classified into three general types: flake (also known as plumbago), crystalline (vein), and amorphous, varying in physical properties, appearance, chemical composition, and impurities. These differences stem from the type of precursor material (oil, coal, or other carbonaceous deposits) and the natural process by which graphite was formed. Table 10.2 summarizes the characteristics of these three types.[2][3]

Table 10.1. Characteristics and Properties of the Three Types of Natural Graphite

Property	Type		
	Flake	Crystalline	Amorphous
Composition			
Carbon, %	90	96	81
Sulfur, %	0.10	0.70	0.10
Density, g/cm^3	2.29	2.26	2.31
Degree of graphitization, %	99.9	100	28
d-spacing (002), nm	0.3355	0.3354	0.3361
Resistivity, $\Omega \cdot$cm	0.031	0.029	0.091
Morphology	Plate	Plate Needle	Granular

1.3   Occurrence and Production

Flake graphite is found disseminated in metamorphosed silica-rich quartzites, gneisses, and marbles. Crystalline (vein) graphite occurs in transverse, igneous, or metamorphic rocks, where it was formed by the transformation of oil precursors. Amorphous graphite is the result of the metamorphosis of coal exposed to high pressure. Graphite is also present in the universe as evidenced by near-perfect crystals frequently found in meteorites.

Besides the U.S., the major producers of graphite are Brazil (flake), Canada (flake), China (flake and amorphous), Korea (amorphous), Madagascar (flake), Mexico (amorphous), Sri Lanka (flake), and the former Soviet Union (flake and amorphous). The total world production was estimated at over 650,000 tons in 1991 at an average price of $1400/ton for flake and $165/ton for amorphous.[2]

1.4   Processing and Applications

After mining, flake and crystalline graphites are milled, usually with ball mills, although ring rollers and jet mills are also used.

A high percentage of the graphite-flake production goes into the fabrication of refractory crucibles for the metal industry. High-quality materials are required such as No. 1 flake which is >90% carbon, and 20 - 90 mesh. Other major applications are found in friction products, lubricating oils and greases, dry-film lubricants, batteries, conductive coatings, electrical brushes, carbon additives, paints, pencil manufacturing, and others.[4]

2.0   CARBON-DERIVED POWDERS AND PARTICLES

Carbon-derived powders and particles comprise a family of synthetic materials, known under the generic term of carbon black, made by burning hydrocarbons in insufficient air. Carbon-black particles are aggregates of graphite microcrystals, each only a few unit cells in size and so small that they are generally not detectable by diffraction techniques. The physical properties of these materials are essentially determined by the nature and extent of their surface areas.

2.1 Carbon Black

The term carbon black is often narrowed to designate the materials produced by two basic methods known as channel and thermal processes.[1] Other carbon blacks are lampblack and acetylene black (see below).

Channel Process. In the channel process, thousands of small flames of natural gas impinge on a cool metallic surface which can be a channel, a roller, or a rotating disk. The carbon black forms on the cool surface and is then exposed to high temperature in air to oxidize the surface of each particle. These particles are in the form of small spheroids. The channel carbon black has the smallest particle size (~ 10 nm), the highest surface area, and the highest volatile content of all carbon blacks.

Thermal Process. In the thermal process, the carbon black is formed by the thermal decomposition of natural gas in the absence of air in a preheated firebrick-lined chamber. The process produces a coarser grade than the channel process with particle size up to 500 nm and lower surface area.

Composition and Properties. Table 10.2 lists the composition and typical properties of carbon black.

Table 10.2. Composition and Properties of Carbon Black

Composition:	
Hydrogen	0.5 - 1.0 %
Nitrogen	0.02 - 0.09 %
Oxygen	2.5 - 7.0 %
Sulfur	0.01 - 0.03 %
CO_2	0.1 - 1.5 %
CO	0.2 - 4.0 %
Carbon	balance
Surface area:	25 - 150 m^2/g
Particle size:	10 - 500 nm
Oil absorption:	0.5 - 1.5 cm^3/g

Applications. A primary use of carbon black is as a filler in rubber to improve the strength, stiffness, hardness, and wear- and heat-resistance. Carbon-black fillers are found in practically every rubber product.

Carbon black is also the base of most black typographical inks. This is a large market since an average of 16 kg of ink, containing 11 - 13% by weight of carbon black, is required for every ton of news-print. Other applications are found in paints, enamels, lacquers, molded carbon and graphite, and many others.

2.2 Lampblack

Lampblack is one of the oldest forms of carbon, first produced by the ancient Chinese who made it by collecting the soot from a burning oil lamp. It was used then (and now) as a black pigment for inks and paints.[1]

Production. Lampblack is produced in multiple furnaces where a preheated feed of coal tar and petroleum oil is burned in a controlled air flow. The particles are calcined in the flue gas to remove excess residual oil and aromatic compounds and, still entrained by the flue gas, are carried in a large chamber. The sudden expansion of the gases causes the particles to settle by gravity.

The feed-to-air ratio in this confined partial combustion is the critical processing factor that controls the characteristics and properties of the resulting lampblack as shown in Table 10.3.

Table 10.3. Characteristics of Lampblack as a Function of Processing

	Oil-to-air ratio	
	Low	High
Air velocity	High	Low
Yield	High	Low
Color Strength	Low	High
Density	High	Low
Blue Undertone	Strong	Weak

Particle Size. The particle size of lampblack averages 100 - 200 nm and, to a great degree, controls the color and oil adsorption as shown in Table 10.4.

Table 10.4. Effect of Particle Size on Properties of Lamblack

	Particle Size	
	Large	Small
Color	Blue	Black
Color Intensity	Low	High
Oil Adsorption	Low	High

Properties. Lampblack is a soft, flocculent, amorphous material with an apparent density that can vary by a factor of twenty, depending on the amount of occluded gases and the shape of the particle. The occluded gases include hydrogen, oxygen, nitrogen, and often sulfur compounds originating from the feed coal tar.

Composition. A typical composition of lampblack (dried at 105°C) is the following:

Carbon: 90.67%

Hydrogen: 1.20%

Nitrogen: traces

Oxygen: 7.41%

Sulfur: 0.66%

2.3 Acetylene Black

Acetylene black is obtained by the thermal decomposition of acetylene (C_2H_2). The gas is piped into a retort preheated at 800°C where it decomposes in the absence of air. The reaction is strongly exothermic and does not require additional heat.[1]

Composition and Properties. Acetylene black is similar to high-grade lampblack but with greater purity, higher liquid-absorption capacity, and higher electrical conductivity. Table 10.5 lists its composition and typical properties.

Applications. The principle applications of acetylene black are in the production of dry cells and as a filler in rubber and plastic materials, particularly if electrical conductivity is required.

Table 10.5. Composition and Properties of Typical Acetylene Black

Composition:	
Polymerization products	<1 %
Carbon	99 %
Ash	0.03 - 0.04 %
Moisture	0.05 - 0.06 %
Density:	2.05 g/cm^3
Apparent density:	0.02 g/cm^3
Surface area:	65 m^2/g
Particle size:	3-130 nm
Undertone:	blue

3.0 INTERCALATED COMPOUNDS AND LUBRICATION

Compounds of graphite are formed when foreign species such as atoms, ions, or molecules are inserted between the layers of the graphite lattice. These compounds can be divided into two general classes with clearly different characteristics: *(a)* the covalent compounds, and *(b)* the intercalation compounds.[5]

3.1 Covalent Graphite Compounds

As the name implies, the covalent graphite compounds have covalent two-electron bonds between the foreign and the carbon atoms. These bonds disrupt the π bonds between layers (see Ch. 3, Sec. 1.2) and the delocalized π electrons are no longer free to move, thereby causing a drastic reduction in the electrical conductivity of the material. This is accompanied by a loss of planarization of the layers which are converted to a puckered structure, somewhat similar to that of diamond (see Ch. 11, Fig. 11.3). The interlayer spacing increases considerably as a result of this insertion and may reach 0.7 nm (vs. 0.3353 nm for graphite). Two covalent graphite compounds are known, namely graphite oxide and graphite fluoride.[6]

Graphite Oxide. The nominal composition of graphite oxide is C_2O with oxygen atoms forming C-O-C bridges in the meta-position. Hydrogen is also present in varying amounts depending on synthesis conditions. The material is obtained by reacting graphite (usually natural-graphite flakes) with a strong oxidizing agent such as potassium permanganate (K_2MnO_4), potassium chlorate ($KClO_3$) or fuming nitric acid (HNO_3).

The structure of graphite oxide is still not clearly defined and several models have been proposed with various hydroxyl-, carbonyl-, ether-bridges and C=C bonds. The interlayer spacing is 0.6 - 0.7 nm. A typical structure is shown in Fig. 10.1.[7] Graphite oxide is strongly hygroscopic and a stronger dehydrating agent than silica gel.

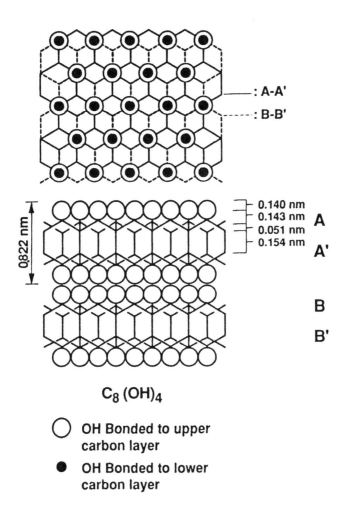

$C_8 (OH)_4$

◯ OH Bonded to upper carbon layer

● OH Bonded to lower carbon layer

Figure 10.1. Proposed model of the ideal structure of $C_8(OH)_4$.[7]

Graphite Fluoride.   The fluorination of graphite (usually natural-graphite flakes) produces a graphite fluoride with a variable composition CF_x (x = 0.3 - 1.1).   A substoichiometric composition is obtained at low fluorination temperature but, under proper conditions, the stoichiometric material, CF, is readily obtained.   When small particles of graphite are fluorinated, composition may reach $CF_{1.12}$ due to the formation of CF_2 groups at the edges of the graphite plates. The interlayer spacing may reach 0.8 nm.   The structure of graphite fluoride is shown in Fig. 10.2.[8]   The x-ray diffraction patterns of graphite oxide, graphite fluoride, and fluorinated graphite oxide are shown in Fig. 10.3.[7]

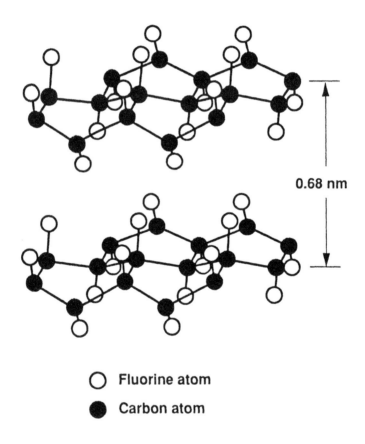

0.68 nm

○   Fluorine atom

●   Carbon atom

Figure 10.2.   Structure of graphite fluoride.[8]

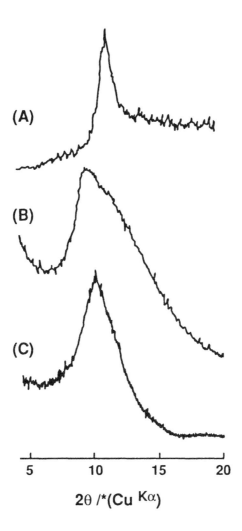

(A) Graphite Oxide (I_c= 0.765 nm)

(B) Fluorinated Graphite Oxide (I_c = 0.920 nm)

(C) Graphite Fluoride (I_c = 0.893 nm)

Figure 10.3. X-ray diffraction patterns of graphite intercalated compounds.[7]

3.2   Graphite Intercalation Compounds

Like the covalent graphite compounds, the intercalation compounds are formed by the insertion of a foreign material into the host lattice. The structure however is different as the bond, instead of being covalent, is a charge-transfer interaction. This electronic interaction results in a considerable increase in electrical conductivity in the *ab* directions.

Stages. Intercalation compounds have a large spread of composition as the percentage of intercalated material changes by regular steps as shown in Fig. 10.4.[9] In the first stage, intercalation reaches a maximum and the material is considered stoichiometric and is known as a first-stage compound.

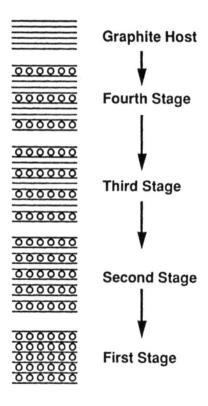

Figure 10.4. The four stagings of the graphite intercalation compound: $C_{12n}K$. Addition of K proceeds through n = 4, 3, 2, and 1.[9]

Donor and Acceptor Compounds. When the intercalated substance donates an electron to the adjacent graphite layer, it is known as a donor, i.e., potassium. When it receives an electron from the layer, it is known as an acceptor, i.e., bromine, arsenic pentafluoride, etc.

Potassium is a common intercalated material. The first stage of a graphite-potassium compound is reached with the limiting formula C_8K, when every carbon layer is separated by a potassium layer. It has the structure shown in Fig. 10.5.[9] Note that all available sites are filled. Upon intercalation, the layers move apart by 0.205 nm, which is less than the diameter of the potassium ion (0.304 nm) indicating that these ions nest within the hexagonal structure of the graphite layer.[9][10]

Many other materials have been intercalated into graphite, including $CoCl_2$ and $NiCl_2$.[11] These materials provide substantial improvement in the tribological properties of graphite.

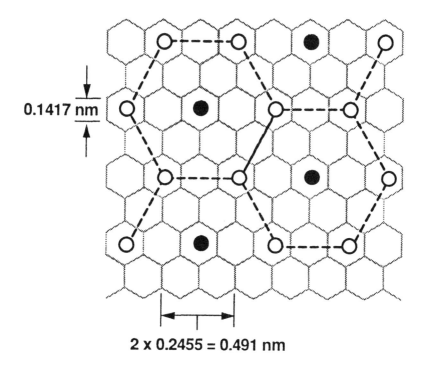

0.1417 nm

2 x 0.2455 = 0.491 nm

Figure 10.5. Structure of potassium-intercalated graphite showing filling of available hexagonal sites in each layer of graphite for: *(a)* the limit of C_8K (K occupies O and ● sites), and *(b)* $C_{12n}K$ (K occupies O sites only)[4]

3.3 Applications

Solid Lubricants. A major application of covalent- and intercalated-graphite compounds is found in solid lubrication. The purpose of solid lubrication is to reduce friction and wear between surfaces in relative motion. The differences between graphite and liquid lubricants are shown in Table 10.6.

Table 10.6. Characteristics of Graphite and Liquid Lubricants

Condition	Graphite	Liquid Lubricants
Vacuum	stable	evaporate
Cryogenic Temperature	stable	freeze
High Pressure	resist load	do not support load
Ionization	stable	decompose
Life	limited	renewable
Thermal Conductivity	low	variable

Graphite and its compounds, molybdenum disulfide (MoS_2), and polytetrafluoroethylene (Teflon™) are the best solid lubricants for most applications. However, they are not suitable in all environments. They perform most effectively when a rolling component to the motion is present, such as pure rolling or mixed rolling/sliding contacts. The efficacy of graphite and graphite fluoride is shown graphically in Fig. 10.6.[8]

Grafoil. Grafoil is usually produced from natural graphite by intercalation with sulfuric or nitric acid, followed by exfoliation by heating rapidly to a high temperature. The resulting flakes are then pressed into a foil which may be subsequently annealed. The foil has low density and an essentially featureless and smooth surface. It has a number of applications such as high-surface materials and high-temperature seals and gaskets.[3]

Electrochemical Applications. As seen above, graphite has the unique ability to intercalate electrochemically positive and negative ions. As such, intercalated graphite has found a number of electrochemical applications, primarily as battery electrodes. An example is a primary battery with high energy-density power based on lithium and fluorine. The anode is lithium and the cathode graphite fluoride. In this particular case, fluorine-intercalated graphite fibers have also been used successfully.[12]

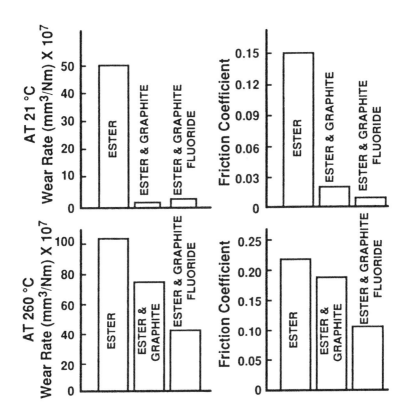

Figure 10.6. Wear and friction of steel with ester lubricant, ester lubricant + 3 weight % graphite, and ester lubricant + 3 weight % graphite-fluoride. Units of wear rate are volume of material removed per unit-load per unit sliding distance.[8]

4.0 ACTIVATION, ADSORPTION AND CATALYSIS

4.1 Charcoal and Activation

Charcoal. Charcoal is a porous form of carbon obtained by the destructive distillation of an organic material in the absence of air. By-products such as wood tar, wood spirit, acetone, and gaseous compounds are usually recovered.[13] A common precursor is coconut shell which produces excellent charcoal. Charcoal is also obtained by heating animal bones and dissolving the calcium phosphate and other mineral compounds with acid. The material is known as "bone black".

Activation. *Activation* is a process that increases the surface area of charcoal and other porous carbon materials. These materials, as produced, have a relatively low porosity. Their structure consists of elementary graphitic crystallites with a large number of free interstices between them. However, these interstices tend to fill with tar-like substances which, on carbonization, block the pore entrances. Opening these pores is accomplished by *activation*.

The Activation Process. *Activation* is essentially a partial oxidation whereby the carbon residues blocking the pores are removed by burning in superheated steam, carbon dioxide, or a combination of the two. Additional increase in porosity may be achieved by further burnoff and by adding activation agents such as $ZnCl_2$, H_3PO_4, KOH, and others.[14] The reactions are the following:

Eq. (1)   with steam: $C + H_2O \rightarrow CO + H_2$ $\Delta H = + 117$ kJ/mol

Eq. (2)   with carbon dioxide: $C + CO_2 \rightarrow 2CO$ $\Delta H = + 150$ kJ/mol

These reactions are endothermic and it is necessary to supply heat to maintain isothermal equilibrium. This is achieved by burning the by-products, either CO or H_2, in situ in air.

The H_2O molecule is smaller than the CO_2 molecule and diffuses faster into the pores of the carbon. Consequently, the reaction rate in Eq. (1) is greater than the rate in Eq (2) and steam activation is the more effective (and the more common) process.

The properties of a typical activated carbon are listed in Table 10.7. In this case, activation was obtained in steam at 57 - 82 kPa for 10 - 12 h.[14]

Table 10.7. Properties of an Activated Peat Semicoke Material

Activation Temp.,°C	C/H Mole Ratio	Weight % Nitrogen	Density g/cm³	F*	d** (nm)
860	13.7	0.48	1.81	0.58	6.5
900	15.9	0.37	1.84	0.62	7.6
1000	28.9	0.30	1.90	0.78	14.1
1040	37.5	0.29	1.91	0.82	18.4

* Ratio micropore volume / micro + mesopore volume
** Diameter of graphitic layers

Activation is now recognized as a simple increase in the internal surface area of the carbon material, resulting in the formation of a well-developed and readily accessible pore structure, with pores of controllable size. The internal surface area of activated carbons ranges from 500 to 1500 m²/g.

4.2 Adsorption

Adsorption can be defined as the formation of a gaseous or liquid layer on the surface of a solid. Because of their unusually large surface area, activated carbons have a high adsorption capability.[15] The ability to adsorb molecules of different sizes is a function of the pore size and can be achieved by controlling the activation process. The micropore size and distribution are expressed by the *Dubinin equation*.[16] Activated carbons are used mostly in the form of granules, although activated fibers such as polyvinyledene chloride (Saran) are also available.[10]

Applications. The applications of activated carbons form a large and growing market and are found in color and odor removal, in water purification, toxic-gas removal, general air purification, metal-ion adsorption for metal recovery, decoloration and purification of sugar, pharmacology, and chromatography.

4.3   Catalyst Support

Catalysts are used on a very large scale in many industrial processes and are an essential part of modern chemical industry. They are characterized by their activity, selectivity, and recycling capability.

A common group of catalysts are the platinum-group metals which have become essential factors in many industrial processes such as gas-phase oxidation, selective hydrogenation of petrochemical and pharmaceutical feedstocks, fuel cells for power generation, and many others. Other common catalysts are iron, nickel, and some transition metals.

These catalysts are in the form of a thin film deposited on a support. The main function of the support is to extend the surface area. However, the support can also alter the rate and the course of the reaction to some degree. The support must be stable at the use temperature and must not react with the solvents, reactants, or by-products.

The two major support materials are activated carbons (commonly called activated charcoal) and activated alumina. Activated carbons impregnated with palladium, platinum, or other metal salts are common in most liquid-phase reactions.

Activated alumina has a lower surface area (75 vs. 350 m^2/g) and is less adsorptive than charcoal. It is also noncombustible (as opposed to charcoal), which is an advantage in regeneration and the burning of carbonaceous residue.

REFERENCES

1. Mantell, C. L., *Carbon and Graphite Handbook*, Interscience Publishers, New York (1968)

2. Kenan, W. M., *Ceramic Bulletin*, 70(5):865-866 (1991)

3. Kavanagh, A. , *Carbon*, 26(1):23-32 (1988)

4. *Tomorrow's Graphite Products Today*, Technical Brochure, Superior Graphite Co., Chicago IL (1991)

5. Boehm, H. P., Setton, R. and Stumpp, E., *Carbon*, 24(2):241-245 (1986)

6. Cotton, F. A. and Wilkinson, G., *Advanced Inorganic Chemistry*, Interscience Publishers, New York (1972)

7. Nakajima, T., Mabuchi, A. and Hagiwara, R., *Carbon*, 26(3):357-361 (1988)

8. Sutor, P., *MRS Bulletin*, 24-30 (May 1991)

9. Huheey, J. E., *Inorganic Chemistry*, 3rd. Ed., Harper and Row, New York (1983)

10. Jenkins, G. M. and Kawamura, K., *Polymeric Carbons, Carbon Fibre, Glass and Char*, Cambridge Univ. Press, Cambridge, UK (1976)

11. Comte, A. A., *ASLE Transactions*, 26(2):200-208 (1983)

12. Dresselhaus, M. S., Desselhaus, G., Sugihara, K., Spain, I. L. and Goldberg, H. A., *Graphite Fibers and Filaments*, Springer-Verlag, Berlin (1988)

13. *Mellor's Modern Inorganic Chemistry*, (G. D. Parkes, ed.), John Wiley & Sons, New York (1967)

14. Wigmans, T., *Carbon*, 27(1):13-22 (1989)

15. Eggers, D. F. et al., *Physical Chemistry*, John Wiley & Sons, New York (1964)

16. Stoekli, H. F., Kraehenbuehl, F., Ballerini, L. and De Bernardini, S., *Carbon*, 27(1):125-128 (1989)

11

Structure and Properties of Diamond and Diamond Polytypes

1.0 INTRODUCTION

The first part of this book deals with graphite and carbon materials, their structure and properties, and their various processes and applications. In this and the next three chapters, the focus is on the other major allotrope of carbon: diamond.

Diamond has outstanding properties, summarized as follows:

- It has the highest thermal conductivity of any solid at room temperature, five times that of copper.

- It is the ideal optical material capable of transmitting light from the far infra-red to the ultraviolet.

- It has an unusually high index of refraction.

- Its semiconductor properties are remarkable, with fifteen times the average electric breakdown of common semiconductors, five times their average hole mobility and a dielectric constant that is half of that of silicon.

- It is extremely resistant to neutron radiation.

- It is by far the hardest-known material.

- It has excellent natural lubricity in air, similar to that of Teflon™.

- It has extremely high strength and rigidity.

- It has the highest atom-number density of any material.

However, diamond is scarce and costly and this has motivated researchers, in the last one hundred years or so, to try to duplicate nature and synthesize it. These efforts are finally succeeding and the scarcity and high cost are now being challenged by the large-scale production of synthetic diamond. The properties of these synthetic diamonds are similar (and in some cases superior) to those of natural diamond at a cost which may eventually be considerably lower.

The Four Categories of Diamond. Modern diamonds belong to one of four distinct categories:

1. Natural diamond, still essentially the only source of gemstones and by far the leader in terms of monetary value (reviewed in Ch. 12).

2. High-pressure synthetic diamond, taking an increasing share of the industrial market, particularly in wear and abrasive applications (reviewed in Ch. 12).

2. CVD (vapor-phase) diamond, potentially important but still basically at the laboratory stage with few applications in production (reviewed in Ch. 13).

4. Diamond-like carbon (DLC), also recent but with growing applications in optics and other areas (reviewed in Ch. 14).

2.0 STRUCTURE OF DIAMOND AND DIAMOND POLYTYPES

2.1 Analytical Techniques

Diamond is often found in combination with other carbon allotropes and it is necessary to clearly identify each material by determining its structure, atomic vibration, and electron state. This is accomplished by the following techniques.

Diffraction Techniques. Diffraction techniques can readily reveal the crystalline structure of bulk diamond or graphite. However, in many cases, a material may be a complex mixture of diamond, graphite, and amorphous constituents on a size scale that makes them difficult to resolve even with electron microscopy and selected area diffraction (SAD). Consequently, the results of these diffraction techniques have to be interpreted cautiously.

In addition, electron diffraction patterns of polycrystalline diamond are similar to those of basal-plane oriented polycrystalline graphite and, when analyzing mixtures of the two, it may be difficult to separate one pattern from the other. Unfortunately, mixed graphite-carbon-diamond aggregates are common in natural and synthetic materials.

Raman Spectroscopy. Fortunately an alternate solution to identification is offered by Raman spectroscopy. This laser-optical technique can determine with great accuracy the bonding states of the carbon atoms (sp^2 for graphite or sp^3 for diamond) by displaying their vibrational properties.[1] The Raman spectra is the result of the inelastic scattering of optical photons by lattice vibration phonons.

As shown in Fig. 11.1, the presence of diamond and/or graphite bonding is unambiguous and clear. Single-crystal diamond is identified by a single sharp Raman peak at 1332 cm (wave numbers), often referred to as the D-band, and graphite by a broader peak near 1570 cm (the G-band) and several second-order features.

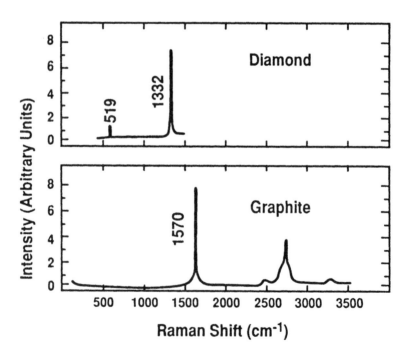

Figure 11.1. Raman spectra of diamond and graphite.

The Raman scattering efficiency for sp^2 bonds is more than fifty times the efficiency for sp^3 bonds for graphitic domains smaller than 10 nm. As a result, the technique is capable of detecting minute amounts of graphite bonds (such as may present in some diamond-like carbon). However, it must be recognized that the techniques cannot readily define the state of aggregation of the constituents.

Diamond Characteristics. It is generally accepted that, for a material to be recognized as diamond, it must have the following characteristics:

- A crystalline morphology visible by electron microscopy
- A single-phase crystalline structure detectable by x-ray or electron diffraction
- A clear diamond Raman spectrum with a sharp peak at 1332 cm

2.2 Atomic Structure of Diamond

In Ch. 2, Sec. 3, the hybridization of the carbon atom from the ground state to the hybrid sp^3 (or tetragonal) orbital state is described. It is shown that this hybridization accounts for the tetrahedral symmetry and the valence state of four with four $2sp^3$ orbitals found in the diamond atomic structure. These orbitals are bonded to the orbitals of four other carbon atoms with a strong covalent bond (i.e., the atoms share a pair of electrons) to form a regular tetrahedron with equal angles to each other of 109° 28', as shown in Fig. 11.2 (see also Fig. 2.10 of Ch. 2).

2.3 Crystal Structures of Diamond

Diamond is a relatively simple substance in the sense that its structure and properties are essentially isotropic, in contrast to the pronounced anisotropy of graphite. However, unlike graphite, it has several crystalline forms and polytypes.

Cubic and Hexagonal Diamond. Each diamond tetrahedron combines with four other tetrahedra to form strongly-bonded, three-dimensional and entirely covalent crystalline structures. Diamond has two such structures, one with a cubic symmetry (the more common and stable) and one with a hexagonal symmetry found in nature as the mineral lonsdaleite (see Sec. 2.5).

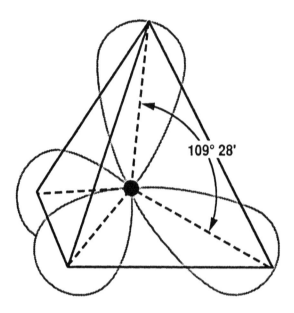

Figure 11.2. The diamond tetrahedron.

Structure of Cubic Diamond. Cubic diamond is by far the more common structure and, in order to simplify the terminology, will be referred to as simply "diamond". The covalent link between the carbon atoms of diamond is characterized by a small bond length (0.154 nm) and a high bond energy of 711 kJ/mol (170 kcal/mol).[2] Each diamond unit cell has eight atoms located as follows: 1/8 x 8 at the corners, 1/2 x 6 at the faces and 4 inside the unit cube. Two representations of the structure are shown in Fig. 11.3, *(a)* and *(b)*.[2][3]

The cubic structure of diamond can be visualized as a stacking of puckered infinite layers (the {111} planes) or as a two face-centered interpenetrating cubic lattices, one with origin at 0,0,0, and the other at 1/4,1/4,1/4, with parallel axes, as shown in Fig. 11.3(c). The stacking sequence of the {111} planes is ABCABC, so that every third layer is identical.

Density of Diamond. With its fourfold coordinated tetrahedral (sp^3) bonds, the diamond structure is isotropic and, except on the (111) plane, is more compact than graphite (with its sp^2 anisotropic structure and wide interlayer spacing). Consequently diamond has higher density than graphite (3.515 g/cm^3 vs. 2.26 g/cm^3).

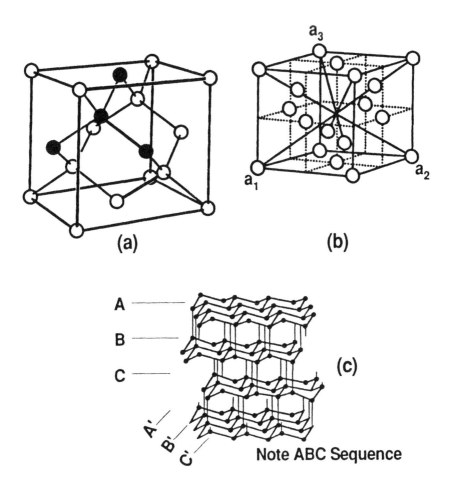

Figure 11.3. Schematics of the structure of cubic diamond.[2][3]

Diamond has the highest atom density of any material with a *molar density* of 0.293 g-atom/cm³. As a result, diamond is the stiffest, hardest and least compressible of all substances. In comparison, the molar density of graphite is 0.188 g-atom/cm³. The atomic and crystal structure data of diamond are summarized in Table 11.1.[4] Also included in the table are the data for hexagonal diamond (see Sec. 2.5).

Table 11.1. Crystal Structure Data of Diamond

Property	Cubic Symmetry	Hexagonal Symmetry
Space group	Fd3m	P6$_3$/mmc
Atoms per unit cell	8	4
Atom position	(000)	(000)
	(1/2 - 1/2 - 0)	(00 - 3/4)
	(0 - 1/2 - 1/2)	(1/8 - 2/3 - 1/2)
	(1/2 - 0 - 1/2)	(1/8 - 2/3 - 7/8)
	(1/4 - 1/4 - 1/4)	
	(3/4 - 3/4 - 1/4)	
	(1/4 - 3/4 - 3/4)	
	(3/4 - 1/4 - 3/4)	
Cell constant at 298 K, nm	0.3567	a = 0.252 c = 0.142
Theoretical density at 298 K, g/cm^3	3.5152	3.52
Carbon-carbon bond distance, nm	0.15445	0.154

2.4   Diamond Crystal Forms

Diamond occurs in several crystal forms (or habits) which include the octahedron, the dodecahedron, and others which are more complicated. As a reminder, the simple crystallographic planes (100, 110 and 111) in a cubic crystal are shown in Fig. 11.4.[5]

These simple planes correspond to the faces of the three major crystal forms of diamond: the {100} cubic, the {110} dodecahedral and the {111} octahedral (Fig. 11.5). Both cubic and octahedral surfaces predominate in high-pressure synthetic diamond where they are found alone or in combination to form blocky crystals.[6]

In CVD diamond, the (111) octahedral and the (100) cubic surfaces predominate and cubo-octahedral crystals combining both of these surfaces are commonly found. Twinning occurs frequently on the (111) surface. Faceted crystals of cut diamonds are predominantly the (111) and (100) surfaces.

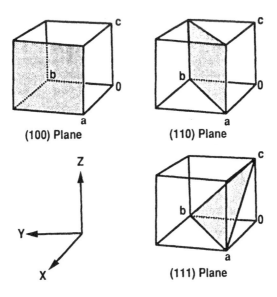

(100) Plane   (110) Plane

(111) Plane

Figure 11.4. Indices of some simple planes in a cubic crystal.[3]

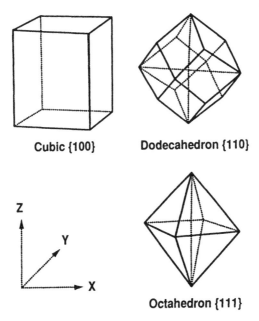

Cubic {100}   Dodecahedron {110}

Octahedron {111}

Figure 11.5. Major crystal forms of diamond.

Diamond Cleavage Planes. Diamond breaks along well-defined cleavage planes (the cleavage of a crystal being its characteristic of breaking along given crystallographic planes where the yield strength is lower due to a high concentration of weaker bonds or a lower total number of bonds). The dominant cleavage plane is the (111) but many others have been observed. This cleavage characteristic is the key to the cutting of gemstones (see Ch. 12). The cleavage energies of the various planes are reviewed in Sec. 6.0.

2.5   The Polytypes of Diamond

Hexagonal Diamond. Hexagonal diamond is an allotropic form of carbon which is close to cubic diamond in structure and properties. It is a polytype of diamond, that is a special form of polymorph where the close-packed layers ({111} for cubic and {100} for hexagonal) are identical but have a different stacking sequence. Hexagonal diamond has an ABAB stacking sequence, so that every second layer is identical as shown in Fig. 11.6.[7] This two-layer hexagonal sequence (known as 2H diamond) is different in this respect from the three-layer sequence of cubic diamond (known as 3C diamond). The crystallographic data of these two polytypes are listed in Table 11.1.

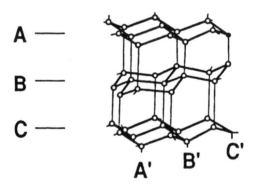

Note ABAB Sequence

Figure 11.6. Schematic of the structure of hexagonal diamond (lonsdaleite).[5]

The cubic diamond nucleus is slightly more stable than the hexagonal with an energy difference between the two of only 0.1 - 0.2 eV per carbon atom. Because of this small energy difference, the growth mechanism leading to the hexagonal structure instead of cubic can readily occur. The inclusion of hexagonal diamond in a cubic diamond structure is equivalent to having a stacking fault at every two-atom layer and is generally detrimental to optical and other properties.

Hexagonal Diamond Occurrence. The formation of hexagonal diamond is usually favored where high carbon supersaturation is prevalent, a condition commonly found during CVD synthesis and occasionally during high-pressure synthesis[8][9] (see Ch. 12). Natural diamond grows at much lower supersaturation levels and consequently natural hexagonal diamond is rarely found.[10] The natural hexagonal diamond is known as the mineral *lonsdaleite.*

Some meteorites contain diamond such as the one found in Canyon Diablo in Arizona. The diamond is in the form of polycrystalline compacts made up of submicron crystals. These crystals are mostly cubic although the hexagonal form is also found.

6H Diamond. Recent investigations have revealed the existence of another intermediate diamond polytype known as 6H diamond.[5] This material is believed to belong to a hypothetical series of diamond types with structures intermediate between hexagonal and cubic. The members of this series are tentatively identified as 4H, 6H, 8H, 10H. The series would include hexagonal (2H) diamond on one end and cubic (3C) diamond on the other (the digit indicates the number of layers). The existence of 4H, 8H and 10H diamonds has yet to be confirmed.

The 6H structure has a mixed six-layer hexagonal/cubic stacking sequence AA´C´B´BC, shown schematically in Fig.11.7.[7] It may exist in CVD diamond.

3.0   IMPURITIES IN DIAMOND AND CLASSIFICATION

3.1   Impurities

The properties of diamond are susceptible to impurities and the presence of even a minute amount of a foreign element such as nitrogen can cause drastic changes. It is therefore important to know the composition of the

A

B

C

A'

B'

C'

A

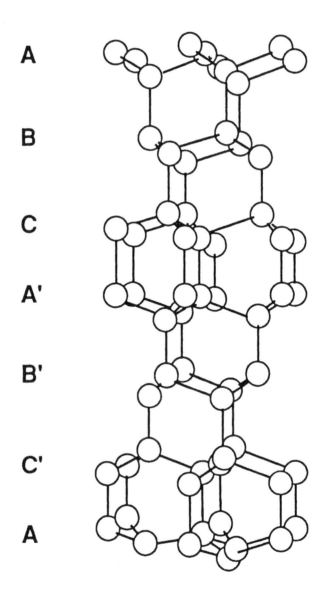

Hexagonal Stacking: A-A', B-B'
Cubic Stacking: Other adjacent Layers

Figure 11.7. Schematic of the structure of 6H diamond.[5]

material being tested as accurately as possible in order to obtain a true evaluation of the measured properties.

Types of Impurities. Diamond, synthetic or natural, is never completely free of impurities. These impurities are divided into two different types:

1. Lattice impurities which consist of foreign elements incorporated in the lattice, the foreign atom replacing a carbon atom.

2. Inclusions, which are separate particles and not part of the lattice, usually consist of silicates of aluminum, magnesium, or calcium such as olivine.[11]

The two major lattice impurities found in diamond are nitrogen and boron. These two elements are the neighbors of carbon in the periodic table. They have small atomic radii and fit readily within the diamond structure. Other elemental impurities may also be present but only in extremely small amounts and their effect on the properties of the material is still uncertain but probably minor.[12]

Nitrogen. Nitrogen impurity in diamond is detected and characterized by IR absorption and paramagnetic resonance. The majority of nitrogen atoms within the diamond structure are arranged in pairs as shown in Fig. 11.8.[11] Isolated nitrogen atoms are rarer. Nitrogen platelets are also present and can be represented as a quasi-planar structure within the cube {100} plane of the diamond crystal. Their exact form is still controversial.

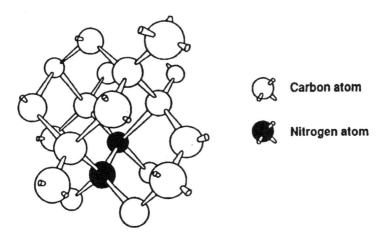

◯ Carbon atom

● Nitrogen atom

Figure 11.8. Schematic of a nitrogen pair impurity in the lattice structure of diamond.[9]

3.2   Classification of Diamonds

No two diamonds have exactly the same composition and properties, and any number of classification schemes can be devised. However only one classification is universally accepted. It is based on the nature and amount of impurities contained within the structure and consists of four types. These types, their origin, and their effect on optical and other properties are summarized in Table 11.2 (some diamonds may consist of more than one type).

Table 11.2. Classification of Diamond

Type	Origin	Impurities
Ia	98% of all natural diamonds	App. 0.1 % nitrogen in small aggregates Includes <10 % platelets Not paramagnetic
Ib	Rare in nature (<0.1%) Includes most high-pressure synthetic diamonds	Nitrogen 0.05 % in lattice Paramagnetic
IIa	Rare in nature	Few ppm of nitrogen Usually clear
IIb	Extremely rare in nature Produced by high-pressure synthesis	Less nitrogen than IIa Becomes semiconductor by boron doping Blue

4.0   PHYSICAL PROPERTIES

4.1   General Considerations

Diamond is costly and available only in small crystals and, as a result, the determination of its properties is difficult and expensive, and the amount of testing and published data are still limited. These problems and the uncertainty about the effect of impurities mentioned above contribute to the considerable spread in the reported values often found in the literature. It is generally agreed that considerable more testing and evaluation are necessary, particularly in the area of synthetic diamond.

The properties listed in this chapter are those of single-crystal diamond, either natural or synthesized at high pressure.

4.2 Thermal Stability

As mentioned above, graphite is the stable allotrope of carbon and is one of the most refractory materials with a sublimation point above 4000 K at one atmosphere. Diamond has a different behavior and is unstable with respect to graphite with a negative free-energy change of 2.88 kJ/mol at room temperature and atmospheric pressure.

Theoretically at least, diamond is not "forever"; graphite would be better qualified. However, in all fairness, the rate of the diamond-graphite conversion is infinitesimally small at ordinary temperatures and, for all practical purposes, diamond is stable, as evidenced by the presence of natural diamonds in some alluvial deposits which were formed over a billion years ago and have not changed since. The carbon phase diagram, illustrated in Fig. 2.20 of Ch. 2, shows the relationship between these two allotropes of carbon.

The free-energy change of the diamond-graphite transition decreases with temperature to reach -10.05 kJ/mol at approximately 1200°C. At that temperature, the transition to graphite is observable but still slow; above it, it proceeds with a rapidly increasing rate as the temperature rises. For instance, a 0.1 carat (0.02 g) octahedral crystal is completely converted to graphite in less than three minutes at 2100°C.[4]

The transformation diamond-graphite is also a function of the environment. It becomes especially rapid in the presence of carbide formers or carbon soluble metals. For instance, in the presence of cobalt, the transformation can occur as low as 500°C. However, in hydrogen diamond is stable up to 2000°C and in a high vacuum up to 1700°C.

The opposite transformation, graphite-diamond, is reviewed in Ch. 12, Sec. 3.3.

5.0 THERMAL PROPERTIES OF DIAMOND

5.1 Summary of Thermal Properties

The thermal properties of diamond are summarized in Table 11.3.

Table 11.3. Thermal Properties of Diamond

Specific heat, C_p, J/mol:	
at 300 K	6.195
at 1800 K	24.7
at 3000 K	26.3
Effective Debye temperature, ΘD:	
273 - 1100 K	1860 ± 10 K
0 K	2220 ± 20 K
Thermal conductivity, W/m·K:	
at 293 K Type Ia	600 - 1000
Type IIa	2000 - 2100
at 80 K Type Ia	2000 - 4000
Type IIa	17,000
Linear thermal expansion, 10^{-6}/K:	
at 193 K	0.4 ± 0.1
at 293 K	0.8 ± 0.1
at 400 - 1200 K	1.5 - 4.8
Standard entropy:	
at 300 K, J/mol·K	2.428
Standard enthalpy of formation:	
at 300 K, J/mol·K	1.884

5.2 Thermal Conductivity

One of the outstanding characteristics of impurity-free diamond its extremely-high thermal conductivity, the highest by far of any solid at room temperature and approximately five times that of copper. This conductivity is similar to that of the graphite crystal in the *ab* direction (see Ch. 3, Table 3.6).

Mechanism of Thermal Conductivity. The thermal conductivity in diamond occurs by lattice vibration. Such a mechanism is characterized by a flow of phonons, unlike the thermal conductivity in metals which occurs by electron transport[13][14] (see Ch. 3, Sec. 4.3, for the mathematical expression of thermal conductivity).

Lattice vibration occurs in diamond when the carbon atoms are excited by a source of energy such as thermal energy. Quantum physics dictates that a discrete amount of energy is required to set off vibrations in a given system. This amount is equal to the frequency of the vibration times the Planck's constant (6.6256 x 10^{-27} erg·s).

Carbon atoms are small and have low mass and, in the diamond structure, are tightly and isotropically bonded to each other. As a result, the quantum energies necessary to make these atoms vibrate is large, which means that their vibrations occur mostly at high frequencies with a maximum of approximately 40 x 10^{12} Hz.[11] Consequently, at ordinary temperatures, few atomic vibrations are present to impede the passage of thermal waves and thermal conductivity is unusually high.

However, the flow of phonons in diamond is not completely free. Several obstacles impede it by scattering the phonons and thus lowering the conductivity.[13] These obstacles include:

- Hexagonal diamond inclusions within the cubic structure and the resulting stacking faults they may create
- Crystallite boundaries, lattice defects, and vacancy sites
- Other phonons (via umklapp processes)
- Point defects due to ^{13}C carbon isotopes, normally 1.1 % of all carbon atoms (See Ch. 2, Sec. 2.1 and Ch. 13, Sec. 3.6)
- Point defects due to impurities

When few of these obstacles are present, diamond is an excellent thermal conductor.

Effect of Impurities on Thermal Conductivity. Of all the obstacles to conductivity listed above, a most important is the presence of impurities, especially substitutional nitrogen. The relationship between thermal conductivity and nitrogen is shown in Fig. 11.9.[15] Nitrogen aggregates, found in Type la crystals, have a much stronger ability to scatter phonons than the lattice nitrogen found in Type lla and lb crystals. The latter contain only a small amount of nitrogen (app. 10^{16} atoms/cm^3); phonon scattering is minimized and these diamond types have the highest thermal conductivity. Other impurities such as boron seem to have much less effect than nitrogen.

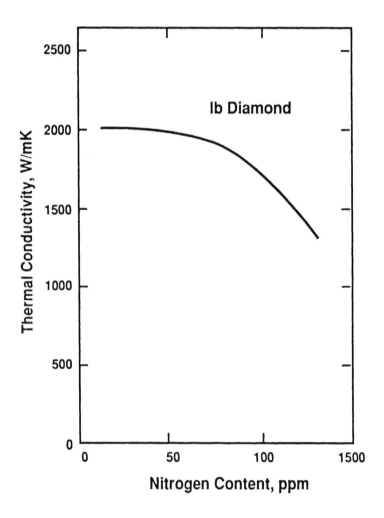

Figure 11.9. Thermal conductivity of Type lb diamond as a function of nitrogen content.[13]

Effect of Temperature on Thermal Conductivity. Fig. 11.10 shows the effect of temperature on the thermal conductivity of several types of diamond.[13]-[15] The conductivity reaches a maximum at approximately 100 K and then gradually drops with increasing temperature. Below 40 K, several materials such as copper have higher conductivity.[14]

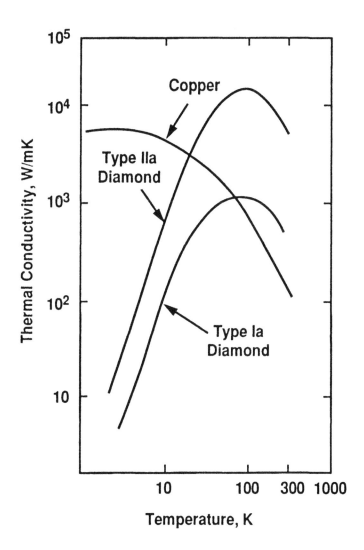

Figure 11.10. Thermal conductivity of Types Ia and II diamonds and copper as function of temperature.[11][12]

5.3 Thermal Expansion

The mechanism of thermal expansion in a crystal material was reviewed in Ch. 3, Sec. 4.4. Like graphite in the *ab* directions, diamond is a strongly bonded solid and, as a result, it has a low thermal expansion. At room temperature, the coefficient of thermal expansion (CTE) is 0.8 ppm·°C (in comparison, copper is 17 ppm·°C and graphite in the *ab* direction is slightly negative). Unlike graphite, diamond has an isotropic thermal expansion which gradually increases with increasing temperature as shown in Table 11.3.

5.4 Specific Heat

The specific heat of diamond is generally comparable to that of graphite and is higher than most metals (see Ch. 3, Sec. 4.3 and Table 3.5). The specific heat of diamond, like that of all elements, increases with temperature (see Table 11.3).

6.0 OPTICAL PROPERTIES OF DIAMOND

6.1 General Considerations

It is now generally accepted that the term "optics" encompasses the generation, propagation, and detection of electromagnetic radiations having wavelengths greater than x-rays and shorter then microwaves, as shown schematically in Fig. 11.11. These radiations comprise the following spectra:

- The visible spectrum which can be detected and identified as colors by the human eye. It extends from 0.4 to 0.7 μm.

- The near-infrared spectrum with wavelengths immediately above the visible (0.7 - 7 μm) and the far infrared (7 μm - ~1 mm). IR radiations are a major source of heat.

- The ultraviolet spectrum with wavelengths immediately below the visible (<0.4 μm). Most UV applications are found in the 0.19 - 0.4 μm range.

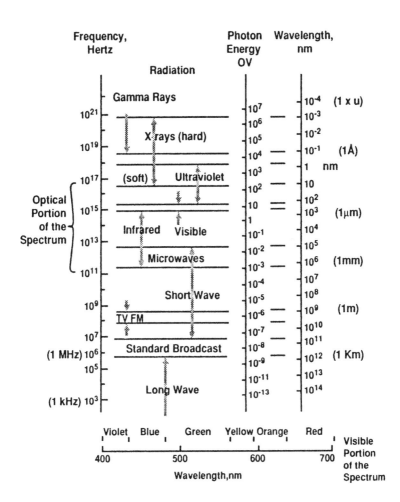

Figure 11.11. The electromagnetic spectrum.

6.2   Transmission

Transmission, or the conduction of radiant energy through a medium, is characterized by transmittance, which is the ratio of radiant power transmitted by a material to the incident radiant power. Transmittance over a wide range of optical wavelengths is one of the optical characteristics of diamond.

Transmission Mechanism. The high transmittance is related to the nature and high strength of the diamond bond. To break these bonds (by exciting an electron across the *bandgap)* requires considerable energy since the bandgap is high (5.48 eV at room temperature).

This excitation can be accomplished by the action of a photon of an electromagnetic radiation. The energy of a photon is proportional to the frequency of the radiation and, as shown in Fig. 11.11, this frequency increases gradually going from the infrared to the visible to the ultraviolet and x-rays. The energy in the lower-frequency radiations such as infrared and visible is too low to excite the diamond electrons across the high bandgap and, as a result, diamond is capable of transmitting across a unusually broad spectral range from the x-ray region to the microwave and millimeter wavelengths and has the widest *electromagnetic bandpass* of any material. In the case of the visible light, virtually none is absorbed and essentially all is transmitted or refracted, giving diamond its unequalled brightness.

Absorption Bands. Pure diamond (which has never been found in nature and has yet to be synthesized) would have only two intrinsic absorption bands as follows:[16][17]

1. At the short wavelength end of the optical spectrum, an ultraviolet absorption due to the electron transition across the bandgap. This corresponds to an absorption edge of 230 nm and, in the ideal crystal, there should be no absorption due to electronic excitation up to that level (Fig. 11.12).

2. An infrared absorption which lies between 1400 and 2350 wave number (cm^{-1}). The IR absorption is related to the creation of phonons and the intrinsic multiphonon absorption. Absorption is nil above $7 \mu m$ (this includes all the major atmospheric windows in the $8 - 14 \mu m$ waveband) (Fig. 11.13).

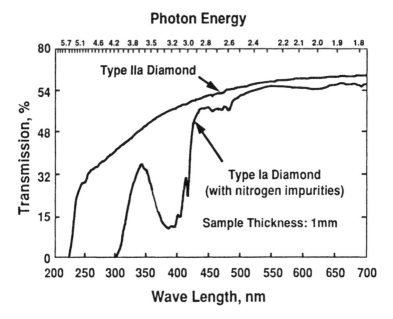

Figure 11.12. Transmission of Types Ia and IIa natural diamonds in the UV and visible spectra.[14]

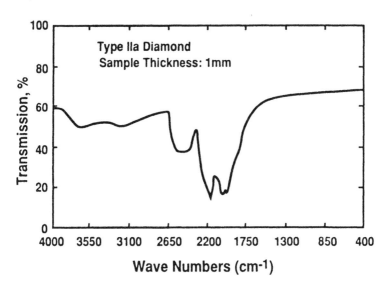

Figure 11.13. Transmission of Type IIa diamond in the infrared spectrum.[14]

Effect of Impurities and Structural Defects. As seen above, diamond would be the ideal transparent material if it were totally free of impurities, particularly nitrogen, and had a perfect structure. However these conditions are never completely achieved, and impurities and crystal-lattice defects and other obstacles to the free movement of photons affect its transmittance. These obstacles add a number of absorption bands to the two mentioned above, particularly in the IR region as shown in Table 11.4. In spite of this, diamond remains the best optical material.

Lattice vacancies (missing atoms) may considerably alter the valence bonds and cause electrons to be exited by a much smaller amount of energy (such as produced by a photon of red light) that would normally be required in a perfect lattice. A diamond containing such lattice vacancies appears blue since the red components of light (the one with less photon energy) are absorbed. A minimum of one vacancy per 10^5 atoms is necessary for the blue color to be noticeable.[11]

Table 11.4. Optical Absorption of Diamond by Type

Type	Optical absorption bands
Ia	IR: 6-13 microns UV: <225 nm
Ib	IR: 6-13 microns UV: <225 nm
IIa	Closer to ideal crystal. No absorption in the range <1332 cm^{-1}. Continuous absorption below 5.4eV
IIb	IR: no significant absorption from 2.5-25 microns UV: absorption at 237 nm

6.3 Luminescence

Visible luminescence is a well-known optical property of single crystal diamond particularly in the blue and green regions. This luminescence originates in the states at mid-bandgap and is caused by impurities and lattice defects. *Cathodoluminescence* (CL) is another characteristic of diamond. The CL of single crystal diamond is described as a band-A luminescence and the peak of the spectra is found between 2.4 and 2.8 eV (from green to purple-blue).[12][18][19]

6.4 Index of Refraction

The *index of refraction* of diamond is high as shown in Table 11.5 and only a few materials have higher indices (Si: 3.5, rutile: 2.9, AlC_3: 2.7, Cu_2O: 2.7). All ionic crystals have lower indices.

Table 11.5. Index of Refraction of Diamond and Selected Materials

Material	Index	Wavelength (nm)
Diamond	2.4237	546.1 (Hg green)
	2.4099	656.28 (C-line)
	2.41726	589.29 (D-line)
	2.43554	486.13 (F-line)
	2.7151	226.5 *
Quartz	1.456	0.656
	1.574	0.185
Crown glass	1.539	0.361
	1.497	2
Air	l.000	0.589

* Near cut-off in the ultraviolet

7.0   X-RAY TRANSMISSION OF DIAMOND

X-ray transmission of diamond is excellent by virtue of its low atomic number and, in thin sections, it even allows the transmission of characteristic x-rays generated by low-energy elements such as boron, carbon, and oxygen.   In this respect, it compares favorably with the standard x-ray window material: beryllium.   The x-ray transmission of a 0.5 mm-thick diamond of the characteristic radiation of a series of elements is shown in Fig.11.14.[20]

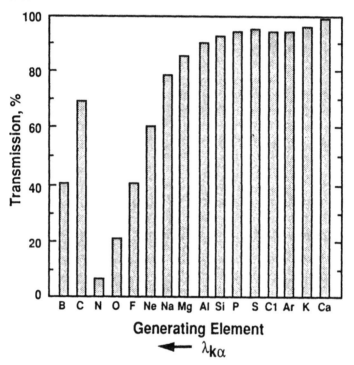

Figure 11.14.   X-ray transmission of a 0.5 mm-thick diamond window of the characteristic radiation of a series of elements.[17]

8.0   ACOUSTICAL PROPERTIES OF DIAMOND

Sound waves are carried by vibrations in the low-frequency range (a few hundred Hertz), unlike thermal conductivity and optical absorption which are associated with high-frequency vibrations.

The structure of diamond favors low-frequency transmission and the material has high sound velocity. Measurements of up to 20 km/s are reported. By comparison, the speed of sound in beryllium is 12.89 km/s and in silicon slightly less than 10 km/s.[11]

9.0 ELECTRICAL AND SEMICONDUCTOR PROPERTIES OF DIAMOND

9.1 Summary of Electrical and Semiconductor Properties

The electrical and semiconductor properties of diamond are summarized in Table 11.6.[21]

Table 11.6. Electrical and Semiconductor Properties of Diamond

Resistivity, $\Omega \cdot m$	
Type I and most Type IIa	10^{18}
Type IIb	$10^3 - 10^5$
Dielectric Constant at 300 K	5.70 ± 0.05
Dielectric Strength, V/cm	10^6
Saturated electron velocity, 10^7 cm/s	2.7
Carrier mobility, $cm^2/V \cdot s$	
Electron	2200
Hole	1600

9.2 Resistivity and Dielectric Constant

Pure single-crystal diamond, with a bandgap of 5.48 eV, is one of the best solid electrical insulators (see Sec. 6.2).[19] The high strength of the electron bond makes it unlikely that an electron would be exited out of the valence band. In pure diamond, resistivity greater than 10^{18} $\Omega\cdot$m has been measured.

However, as with the optical properties, the presence of impurities can drastically alter its electronic state and the inclusion of sp^2 (graphite) bonds will considerably decrease the resistivity and render the material useless for electronic applications.

The *dielectric constant* of diamond (5.7) is low compared to that of other semiconductors such as silicon or germanium but not as low as most organic polymers (in the 2 to 4 range) or glasses (approximately 4).

9.3 Semiconductor Diamond

The semiconductor properties of diamond are excellent and it has good potential as a semiconductor material.[4] It is an *indirect bandgap* semiconductor and has the widest bandgap of any semiconductor (see Sec. 6.2).

When a semiconductor material is heated, the probability of electron transfer from the valence band to the conduction band becomes greater due to the thermal excitation and, above a certain limiting temperature, the material no longer functions as a semiconductor. Obviously the larger the bandgap, the smaller the possibility of electron transfer and large-bandgap semiconductors can be used at higher temperature. This is the case for diamond which has an upper limit semiconductor temperature of 500°C or higher. In comparison, the upper limit of silicon is 150°C and that of GaAS is 250°C.

Diamond can be changed from an intrinsic to an extrinsic semiconductor at room temperature by doping with other elements such as boron and phosphorus.[4][22] This doping can be accomplished during the synthesis of diamond either by high pressure or especially by CVD (see Ch. 13, Sec. 4.4). Doped natural diamond is also found (Type IIb) but is rare.

Diamond has an excellent electron-carrier mobility exceeded only by germanium in the p-type and by gallium arsenide in the n-type. The saturated carrier velocity, that is, the velocity at which electrons move in high electric fields, is higher than silicon, gallium arsenide, or silicon carbide and, unlike other semiconductors, this velocity maintains its high rate in high-intensity fields as shown in Fig. 11.15.

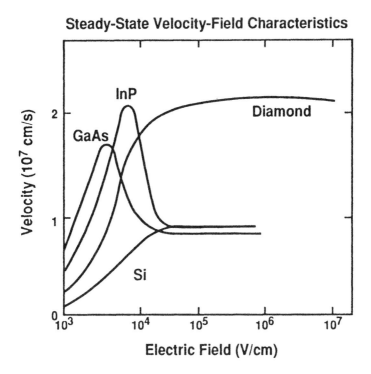

Figure 11.15. Electron-carrier mobility of diamond and other semiconductor materials.

10.0 MECHANICAL PROPERTIES OF DIAMOND

10.1 Summary of Mechanical Properties

It should be emphasized that the strength properties are difficult and expensive to measure due to the lack of diamond of the required test size and configuration. The data presently available show a material of considerable strength and rigidity.

The mechanical properties of diamond are summarized in Table 11.7.[23] The properties were measured on single-crystal diamond either natural or produced by high pressure. For comparison purposes, the table includes the properties of a high-strength ceramic, namely alumina.

Table 11.7. Mechanical Properties of Diamond and Alumina at 23°C

Property	Diamond	Alumina
Density, g/cm^3	3.52	3.98
Young's modulus, GPa	910 - 1250	380 - 400
Compression strength, GPa	8.68 - 16.53	2.75
Knoop hardness, kg/mm^2		
overall	5700 - 10400	2000 - 2100
111 plane	7500 - 10400	
100 plane	6900 - 9600	
Poisson's ratio	0.10 - 0.16	0.22
Coefficient of friction		
in air	0.05 - 0.1	
in vacuum	near 1	

10.2 Hardness

The very fact that diamond is the hardest known material makes it difficult to measure its hardness since only another diamond can be used as an indenter. This may explain the wide variations in reported values which range from 5,700 to over 10,400 kg/mm^2.

The hardness of diamond is compared with that of other hard materials in Fig.11.16. The test method is the Knoop hardness test which is considered the most accurate for crystalline materials. The hardness is also a function of the crystal orientation as shown in Table 11.7.

The hardness of diamond can also be determined form the elastic coefficients as there is a linear relation between hardness and these coefficients.

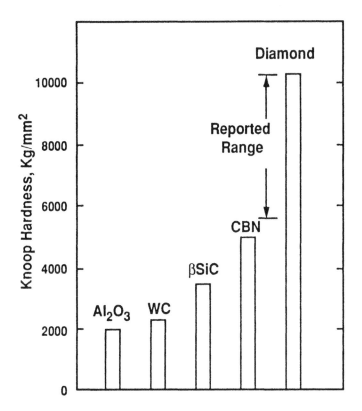

Figure 11.16. Hardness of diamond and other hard materials.[17]

10.3 Cleavage Planes

Diamond behaves as a brittle solid and fractures readily along its cleavage planes. Cleavage occurs mostly along the {111} planes but can also occur along the other planes since the energy differences between planes is small as shown in Table 11.8.[23] Cleavage velocity in all these planes is considerable and has been measured at several thousands of meters per second.

Table 11.8. Theoretical Cleavage Energies of Diamond

Plane	Angle between plane and (111) plane	Cleavage energy $(J \cdot m^{-2})$
111	0° and 70° 32'	10.6
332	10° 0'	11.7
221	15° 48'	12.2
331	22° 0'	12.6
110	35° 16' and 90°	13.0
100	54° 44'	18.4

10.4 Friction

Measured in air, diamond has one of the lowest *coefficients of friction* of any solid. This low friction however is a surface property which is apparently dependent on the presence of oxygen and other adsorbed impurities. In high vacuum, the chemisorbed species are removed and the coefficient of friction increases considerably and approaches one.[24]

11.0 CHEMICAL PROPERTIES OF DIAMOND

Much of the information on the chemical properties of diamond refers to the single-crystal material, either natural or high-pressure synthesized and, in some cases, to polycrystalline films. In the latter case, it is expected that grain boundary, structure, and the concentration of impurities at the boundaries play a role in controlling the chemical properties.

11.1 Oxidation

Diamond is generally inert to most chemical environments with the notable exception of oxidation. In pure oxygen, the onset of oxidation has been shown to start at temperature as low as 250°C for finely divided powders and to become rapid above 600°C. Diamond burns brightly in an oxygen jet at 720°C. The reaction is as follows:

$$C(\text{diamond}) + O_2 \rightarrow CO_2 \ (g)$$

The by-product of the reaction is carbon dioxide which, being a gas, provides no surface passivation. A fresh diamond surface is always exposed and oxidation proceeds parabolically with temperature. Oxidation in air is less rapid with the onset of the reaction at approximately 500°C.

Under normal conditions, oxygen is adsorbed on the surface of diamond after exposure to air (or oxygen) for a period of time. However, no adsorption occurs if the temperature is below -78°C. From 0 - 144°C, oxygen is chemisorbed. CO_2 is formed from 244 to 370°C by the interaction of O_2 with the diamond surface.[24] Adsorbed oxygen and carbon oxides account for the hydrophillic characteristic of diamond. As mentioned in Sec. 10.4, the formation of surface oxides is also an important factor in the control of frictional properties.

At 600°C and low pressure, the presence of residual oxygen results in the formation of a dense film of graphite.

11.2 Reaction with Hydrogen

Like oxygen, hydrogen is chemisorbed on the surface of diamond but not until a temperature of 400°C is reached. This chemisorption is probably the result of the formation of surface hydrides. Diamond is generally considered inert to molecular hydrogen (as opposed to graphite). However attack by atomic hydrogen occurs above 1000°C. Yet diamond is far less reactive than graphite, a characteristic which is used to good advantage in the deposition of diamond films and the selective elimination of the co-deposited graphite (see Ch. 13).

11.3 General Chemical Reactions

Diamond is resistant to all liquid organic and inorganic acids at room temperature. However it can be etched by several compounds including strong oxidizers such as sodium and potassium nitrates above 500°C, by fluxes of sodium and potassium chlorates, and by molten hydroxides such as NaOH. It is resistant to alkalis and solvents. At approximately 1000°C, it reacts readily with carbide-forming metals such as Fe, Co, Ni, Al, Ta, and B. This characteristic provides the mechanism of high-pressure synthesis (see Ch. 12, Sec. 4.5). Generally speaking, diamond can be considered as one of the most-chemically resistant material.

REFERENCES

1. Nemanich, R. J., *J. Vac. Sci. Technol.*, A6(3):1783-1788 (May/June 1988)

2. Guy, A. G., *Elements of Physical Metallurgy*, Addison-Wesley Publishing, Reading, MA (1959)

3. Cullity, B. D., *Elements of X-Ray Diffraction*, Addison-Wesley Publishing, Reading, MA (1956)

4. Spear, K. E., *J. Am. Ceram. Soc.*, 72(2):171-191 (1989)

5. Eggers, D. F., Jr. and Halsey, G. D., Jr., *Physical Chemistry*, John Wiley & Sons, New York (1964)

6. Gardinier, C. F., *Ceramic Bulletin,* 67(6):1006-1009 (1988)

7. Spear, K. E., Phelps, A. W. and White, W. B., *J. Mater. Res.,* 5(11):2271-85 (Nov. 1990)

8. Dawson, J. B., *The Properties of Diamond,* (J. E. Field, ed.), 539-554, Academic Press, London (1979)

9. Bundy, F. P. and Kasper, J. S., *J. Chemical Physics*, 46(9) (1967)

10. Angus, J. C., *Diamond Optics*, 969:2-13, SPIE, (1988)

11. Davies, G., *Diamond,* Adam Hilger Ltd., Bristol, UK (1984)

12. Sellschrop, J. P., *The Properties of Diamond,* (J. E. Field, ed.), 108-163, Academic Press, London (1979)

13. Berman, R., *Properties of Diamond,* (J. E. Field, ed.), 3-23, Academic Press, London (1979)

14. Singer, S., *Diamond Optics*, 969:168-177, SPIE, (1988)

15. Yazu S., Sato, S. and Fujimori, N., *Diamond Optics*, 969:117-123, SPIE (1988)

16. Seal M. and van Enckevort, W., *Diamond Optics,* (A. Feldman and S. Holly, eds.), 69:144-152, SPIE (1988)

17. Lettington, A. H., *Applications of Diamond Films and Related Materials*, (Y. Tzeng, et al, eds.), 703-710, Elsevier Science Publishers (1991)

18. Kawarada, H., *Jpn. J. of App. Physics*, 27(4):683-6 (Apr. 1988)

19. Davies, G., *Diamond Optics,* (A. Feldman and S. Holly, eds.), 969:165-184, SPIE (1988)

20. Conner, L., CVD Diamond—Beyond the Laboratory, in *Proc. Conf. on High-Performance Thin Films*, GAMI, Gorham, ME (1988)

21. Collins, A. T. and Lightowlers, E. C., *Properties of Diamond*, (J. E. Field, ed.), 79-106, Academic Press, London (1979)

22. Fujimori, N., *New Diamond*, 2(2):10-15 (1988)

23. Field, J. E., *The Properties of Diamond*, (J. E. Field, ed.), 282-324, Academic Press, London (1979)

24. Boehm, H. P., *Advances in Catalysis*, 179-272, Academic Press, New York (1966)

12

Natural and High-Pressure Synthetic Diamond

1.0 INTRODUCTION

Rough diamonds, that is, uncut and unpolished, were known and prized in antiquity. They were first reported in India 2700 years ago. From India, diamond trading moved gradually westward through Persia and the Roman Empire. However the full beauty of diamond was not uncovered until faceting and polishing techniques were developed in the 14th and 15th centuries. A detailed history of diamond is given in Refs. 1 and 2.

Unlike graphite and carbon materials, diamond is very rare and, with opal and ruby, considered the most valuable mineral, known the world over as a gemstone of perfect clarity, brilliance, hardness, and permanence.

Diamond is produced in nature at high pressure and temperature in volcanic shafts. The high-pressure synthesis essentially duplicates this natural process and both materials, the natural and the synthetic, have similar properties and are reviewed together in this chapter.

2.0 NATURAL DIAMOND

2.1 Occurrence and Formation of Natural Diamond

The two major allotropes of the element carbon, graphite and diamond, occur in igneous rocks.[3] As seen in Ch. 11, at ordinary pressures graphite is the stable form at all temperatures while diamond is theoretically stable only at high pressures. These pressures are found deep within or under the earth's crust as a result of the weight of overlying rocks. Diamond is formed by crystallization from a carbon source if temperature is sufficiently high. In order to retain its structure and avoid being transformed into graphite by the high temperature, diamond must be cooled while still under pressure. This would occur if it is moved rapidly upward through the earth's crust. A rapid ascent is also necessary to minimize any possible reaction with the surrounding, corrosive, molten rocks.

Such circumstances were found during the formation of some *ultramafic* bodies as evidenced by their pipe-like form and *breccia*-like structure (i.e., with large angular fragments), indicating a rapid upward motion. The composition of the transporting liquid and especially the presence of oxidizing agents such as carbon dioxide and water were such that corrosion was minimized and the diamond crystals were preserved.

Source of Carbon in Igneous Rocks. The source of carbon in *igneous* rocks is still controversial. It could be an original constituent in materials deep in the crust or mantle, or it could be organic materials from partially-melted sedimentary rocks or carbonates.

Diamond Minerals. The mineral *kimberlite* is so far the major source of natural diamond. New information and new studies in progress, particularly in Russia, may add evidence of additional origins for diamond besides kimberlite magma.[2][4][5]

Interstellar Diamond. Diamond has also been found in meteorites and has been detected in dust generated by supernovas and red giants (see Ch. 11, Sec. 2.5).

2.2   Processing of Natural Diamond

Kimberlite is the principal diamond bearing ore. In a typical mine such as the Premier Mine near Pretoria, South Africa, one hundred tons of kimberlite produce an average of thirty-two carats of diamond (6.4 g). Diamonds are sorted from the mineral by an x-ray beam; the diamond luminesces with the x-ray and the luminescence activates an air jet which propels the diamond into a separate bin (Fig. 12.1).[2] Gemstones (a very small percentage) are then separated from the industrial-quality material.

In the grading of diamond for industrial purposes, suitable whole stones are selected to be cleaned, cleaved, sawed, ground, drilled, or metal-coated to achieve the desired shapes and bonding characteristics for applications such as well-drilling tools and dressers. Lower-quality stones and crushing *bort* are processed with hammer and ball mills to achieve the desired particle sizes for other applications such as grinding wheels and lapping compounds.[6]

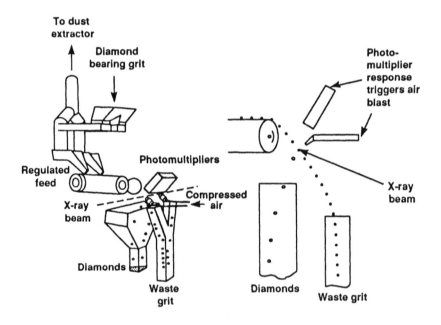

Figure 12.1. Schematic of x-ray diamond sorter.[2]

Diamond Cutting. A rough diamond must be cut to obtain the optimum shape and best polishing faces. Diamond cutting requires a thorough knowledge of the crystallography and many years of practice (see Ch. 11, Sec. 2.4). Cutting a diamond results in an weight loss of 50% or more, depending on the cut. For instance, to obtain a one carat brilliant requires a 2 to 2.5 carat octahedron. The cost of cutting is, of course, reflected in the final cost which can be five or six times that of the rough diamond.

2.3   Characteristics and Properties of Natural Diamond

Gemstones are identified by the following characteristics (known as the four C's).

- Carat: the weight of the stone (1 carat = 0.2 g). The carat is divided into points (100 points to the carat) and a typical stone weight is 8 points.

- Cut: the quality of shape and polishing. Cuts can be pear, emerald, marquise, or brilliant (58 faces) and are designed to enhance refraction and brilliance.

- Clarity: a flawless diamond has no visible imperfection under a 10-power loupe. A flawed diamond has imperfection detectable by the naked eye.

- Color: colorless diamond are the most valuable. The so-called "fancy colors", red, green, and blue, are also in great demand.

Unprocessed natural diamond has a surface that can be brilliant (adamantine), frosted or dull. It comes in many colors from black to essentially colorless. These colors are caused by impurities or by defects in the crystal lattice and, among gemstones, the most common are pale yellow, pale green, pale blue, and pink. Pale blue is the most valuable and is the color of the finest gemstones such as the famous Hope diamond.[2]

Natural diamond is divided in four types based on optical and other physical characteristics and usually derived from the amount and distribution of nitrogen within the crystal lattice. These types are described in Ch. 11, Sec. 3.1 and 3.2, and Table 11.2.

A relatively rare form of natural diamond, found mostly in Brazil, is called *carbonado*. It is a polycrystalline aggregate containing graphite and other impurities. It is much tougher than the single crystal and has found a niche in specific grinding applications such as drill crowns which require a tough material. A similar structure is now obtained by high-pressure synthesis (see Sec. 5.4)·

The physical and chemical properties of natural diamond are generally similar to the properties of the single crystal reviewed in Ch. 11.

3.0   HIGH-PRESSURE SYNTHETIC DIAMOND

3.1   Historical Review

In 1814, the English chemist H. Davy proved conclusively that diamond was a crystalline form of carbon. He showed that only CO_2 was produced when burning diamond without the formation of aqueous vapor, indicating that it was free of hydrogen and water. Since that time, many attempts were made to synthesize diamond by trying to duplicate nature. These attempts, spread over a century, were unsuccessful (some bordering on the fraudulent). It was not until 1955 that the first unquestioned synthesis was achieved both in the U.S. (General Electric), in Sweden (AESA), and in the Soviet Union (Institute for High-Pressure Physics). Table 12.1 summarizes these historical developments.[1][2][5][7]

Table 12.1. Historical Development of High-Pressure Synthetic Diamond

1814	Carbon nature of diamond demonstrated by Davy
1880	Sealed-tube experiments of Harvey
1894	Carbon-arc experiments of Moissant
1920	Unsuccessful synthesis attempts by Parson
1943	Inconclusive synthesis experiments of Gunther
1955	First successful solvent-catalyst synthesis by General Electric, AESA, Sweden, and in the Soviet Union
1957	Commercial production of grit by General Electric
1965	Successful shock-wave experiments by Dupont
1983	Production of a six-carat stone by de Beers
1990	Commercial production of 1.4 carat stones by Sumitomo

3.2   The Graphite-Diamond Transformation

The stability of graphite and diamond and the diamond-graphite transformation were reviewed in Ch. 11, Sec. 4.2. This transformation is mostly of academic interest since few people would want to obtain graphite from diamond. However the opposite transformation, graphite to diamond, is of considerable importance.

Graphite transforms into diamond upon the application of pressure P (in atm) and temperature T (K). This relationship is expressed by the following equation:

Eq. (1) $P^{eq/atm} = 7000 + 27T$ (for T>1200 K)

The equation was determined from extensive thermodynamic data which include the heat of formation of graphite-diamond, the heat capacity of graphite as a function of temperature, and the atomic volume and coefficient of thermal expansion of diamond. Some of these data are still uncertain and the generally-accepted values are listed in Table 12.2.

The PT relationship of Eq. 1 is shown graphically in Fig. 12.2.[7] It has been generally confirmed by many experiments.

Table 12.2. Characteristics of the Transition Reaction of Graphite-Diamond[8]

ΔH°_{298}, J/mol	1872 +/- 75
ΔS°_{298}, J/mol·K	-3.22
ΔC_p above 1100 K, J/mol·K	0
Equilibrium pressure at 2000 K, Pa	64×10^8
Volume change at 2000 K transition, cm³/mol	1.4
Atomic volume, cm³/mol	
$\quad$ V graphite	5.34
$\quad$ V diamond	3.41
$\quad$ ΔV	- 1.93

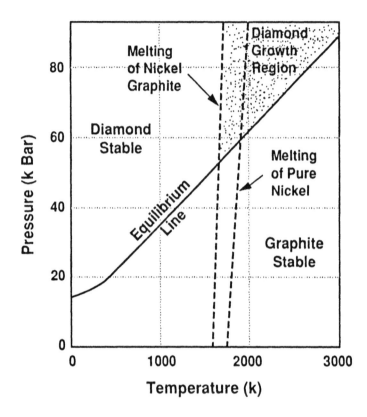

Figure 12.2. Pressure-temperature diagram of diamond-graphite with melting lines of nickel and nickel-graphite eutectic.[7]

The Kinetic Barrier. Although thermodynamically feasible at relatively low pressure and temperature, the transformation graphite-diamond faces a considerable kinetic barrier since the rate of transformation apparently decreases with increasing pressure. This kinetic consideration supersedes the favorable thermodynamic conditions and it was found experimentally that very high pressure and temperature (>130 kb and >3300 K) were necessary in order for the direct graphite-diamond transformation to proceed at any observable rate.[7][9] These conditions are very difficult and costly to achieve. Fortunately, it is possible to bypass this kinetic barrier by the solvent-catalyst reaction.

3.3   Solvent-Catalyst High-Pressure Synthesis

Solvent-Catalyst Reaction. The solvent-catalyst process was developed by General Electric and others. It establishes a reaction path with lower activation energy than that of the direct transformation. This permits a faster transformation under more benign conditions. As a result, solvent-catalyst synthesis is readily accomplished and is now a viable and successful industrial process.

Not all carbon materials are suitable for the solvent-catalyst transformation. For instance, while graphitized pitch cokes form diamond readily, no transformation is observed with turbostratic carbon.[10]

The solvent-catalysts are the transition metals such as iron, cobalt, chromium, nickel, platinum, and palladium. These metal-solvents dissolve carbon extensively, break the bonds between groups of carbon atoms and between individual atoms, and transport the carbon to the growing diamond surface.

The solvent action of nickel is shown in Fig. 12.2. When a nickel-graphite mixture is held at the temperature and pressure found in the cross-hatched area, the transformation graphite-diamond will occur. The calculated nickel-carbon phase diagram at 65 kbar is shown in Fig. 12.3. Other elemental solvents are iron and cobalt.[7][11][12] However, the most common catalysts at the present time are Fe-Ni (Invar™) and Co-Fe. The pure metals are now rarely used.

The Hydraulic Process. The required pressure is obtained in a hydraulic press shown schematically in Fig. 12.4.[7] Pressure is applied with tungsten-carbide pistons (55 - 60 kb) and the cell is electrically heated so that the nickel melts at the graphite interface where diamond crystals begin to nucleate. A thin film of nickel separates the diamond and the graphite as the diamond crystals grow and the graphite is gradually depleted.

The hydraulic process is currently producing commercial diamonds up to 6 mm, weighing 2 carats (0.4 g) in hydraulic presses such as the one shown in Fig. 12.5. Micron-size crystals are produced in a few minutes; producing a two-carat crystal may take several weeks. Typical crystals are shown in Fig. 12.6. Even larger crystals, up to 17 mm, have recently been announced by de Beers in South Africa and others. Research in high-pressure synthesis is continuing unabated in an effort to lower production costs and produce still-larger crystals.

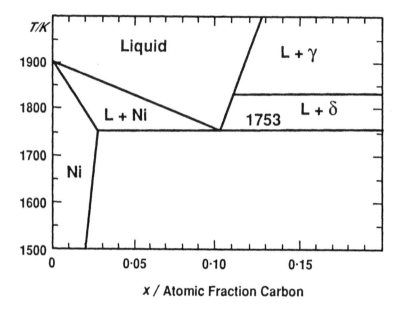

Figure 12.3. Nickel-carbon phase diagram at 65 kbar:[11] α = graphite, δ = diamond

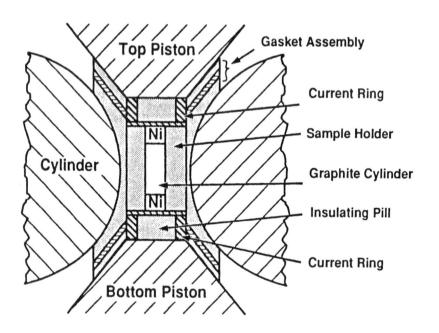

Figure 12.4. High-pressure cell for the production of diamond. Note direct heating of nickel-graphite-nickel cylinder.[7]

Figure 12.5.   High-pressure diamond press.   *(Photograph courtesy of GE Superabrasives, Worthington, OH.)*

Figure 12.6. Typical high-pressure synthetic diamond crystals. *(Photograph courtesy of GE Superabrasives, Worthington, OH.)*

3.4 Shock-Wave Processing

The high pressure generated by the shock waves from an explosion can produce the direct and essentially immediate conversion of graphite into diamond.[13]

A schematic of the process is shown in Fig. 12.7. A mixture of graphite and nodular iron is placed inside a 25 cm diameter cavity in a lead block. A flat metal plate, uniformly coated with TNT on the back side, is placed in front of the cavity. The TNT is detonated and the plate impacts the cavity at a peak velocity of ~ 5 km/s. A peak pressure estimated at 300 kbar and a temperature of approximately 1000 K are maintained for a few microseconds. The formation of diamond is assisted by the presence of an iron solvent-catalyst. The diamond crystals are then separated by selective acid digestion and sedimentation.

Only very small polycrystalline diamonds are produced with a maximum particle size of approximately 60 μm. The technique is commercialized by Dupont in the U.S. Because of its small particle size the material is limited to applications such as polishing compounds.

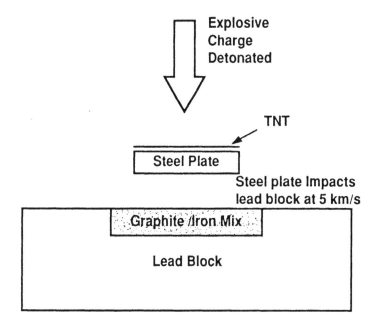

Figure 12.7. Schematic of shock-wave processing of diamond.

4.0 NATURAL AND HIGH-PRESSURE SYNTHETIC DIAMOND PRODUCTION

4.1 Introduction

Until World War II, the diamond business was relatively simple, consisting only of natural diamond. The great majority of the market was gemstones and industrial applications were limited. The advent of synthetic diamond and the rapid rise of industrial applications have drastically altered the industry, and more changes will undoubtedly take place in the future such as the development of CVD diamond (Ch. 13). The challenge and potential impact of this new technology are well understood by the diamond producers.

4.2 Gemstones and Industrial Diamond

The diamond business is divided into two major categories: gemstones and industrial diamond.[14][15] Production and market values are shown in Table 12.3.

The gemstone business comprises 93% of the diamond business in monetary terms but only approximately 1% on a weight basis, reflecting the very large cost difference between the two categories. The average cost of industrial diamond is $1.10/carat and that of gemstone is >$1000/carat and, in some extreme cases, may reach $60,000/carat for blue diamond Type II. The price of natural diamond has remained relatively steady in the last few years.

Table 12.3. The Estimated Annual Diamond Business in 1992

	Gemstone	Industrial Diamond
Production (tons)	~ 1 (all natural)	90 70 (synthetic) 20 (natural)
Market value	$7 billion	$500 million

4.3 Production of Natural Diamond

The worldwide production of natural diamond has risen steadily since World War II, in answer to increasing demand (mostly for gemstones) and as a result of improved prospecting and mining techniques and the opening of new mines. Production, which was only two tons in 1947, reached an estimated twenty tons in 1992.

The major diamond-producing countries in 1990 are shown in Table 12.4.[14][15]

The tonnage production figures of Table 12.4 do not reflect the monetary value. Australia produces a third of the world tonnage but very few gemstones (such as the "fancy pink" for which it is famous). Nanibia's production is less than half the tonnage but more than twice the value of that of Australia.

Table 12.4. Natural Diamond Producing Countries in 1990

Country	Yearly Production	Comments
Australia	7 tons	<5% gem quality
W. & Central Africa	6 tons	nearly all industrial grade
Botswana	3 tons	1/3 gems, largest gem producer
Russia	2 tons	gem and industrial, export only, domestic use unknown
South Africa	1.5 tons	gem and industrial production declining
Nanibia (ex SW Africa)	small	alluvial fields with high % gemstones

4.4   The Diamond Gemstone Market

As mentioned above, diamond gemstones still remain the major use of diamond in monetary terms in a market tightly controlled by a worldwide cartel dominated by the de Beers organization of South Africa.[1] Some maintain that the cartel keeps the cost of gemstones at an artificially high level and that the supply is abundant. Control of the price of industrial diamond is not quite as obvious.

The competition from high-pressure synthetic gemstones is a possibility. Gem-quality synthetic crystals up to fifteen carats are regularly produced but at this time they are at least as expensive as the natural ones.

A special case is blue diamond which is obtained by adding boron to the metal catalyst and could be sold profitably in competition with the rare and expensive natural stones. But synthetics face a psychological barrier from the very fact that they are synthetic, although indistinguishable from the natural.[16] Also the tight rein of the cartel and the possibility of dumping is always present.

4.5   High-Pressure Synthetic Diamond Production

The major producers of high-pressure synthetic diamond include General Electric in the U.S., de Beers in South Africa, Ireland and Sweden, Sumitomo and Tome in Japan, and plants in the former Soviet Union, Czechoslovakia, Germany, Korea, and China.

The growth of high-pressure synthetic diamond is shown in Fig. 12.8.[14] The market is now of considerable size with an estimated production of seventy tons in 1992, all for industrial applications. The largest consumer is the U.S., closely followed by Japan and Western Europe.

High-pressure synthetic diamond, because of its high purity and uniformity, has taken an increasing share of the industrial diamond market and has replaced natural diamond in many areas.

5.0   INDUSTRIAL APPLICATIONS OF NATURAL AND HIGH-PRESSURE SYNTHETIC DIAMONDS

Grinding, grooving, cutting, sharpening, etching, and polishing are the main applications of industrial diamond and consume approximately 90% of the production (natural and synthetic) in the processing of stone, metal, ceramic, glass, concrete, and other products. Other applications in optics and other areas have a more limited but growing market.

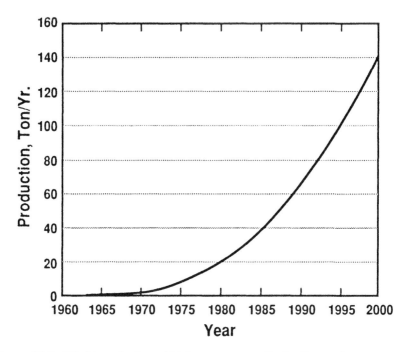

Figure 12.8. World production of high-pressure synthetic diamond.

5.1 Industrial Diamond Powder, Grit, and Stones

Small synthetic crystals (<1 mm) are used as powder and grit (in six different sizes), in the manufacture of fine grinding, lapping and polishing compounds, matrix-set drill bits and tools, saw blades, and polycrystalline diamond (PCD) (see Sec. 5.2).

Larger crystals (>1 mm) are polished and shaped and handled as individual stones in the manufacture of drill bits, glass cutters, abrasive wheel truers, single-point turning, shaping and cutting tools, heavy-duty truing and large plunge cutting tools, and wire-drawing dies. The main advantages of the synthetic stones over the natural are the ability to control uniformity, size, crystal habit, crystal friability, and generally higher quality and consistency.[6]

Natural industrial diamond is now limited to the low-grade (and cheaper) applications such as nail files, low-cost saw blades, and polishing compounds.

5.2 Diamond Cutting and Grinding Tools

Cutting and Grinding Tools. Cutting tools have a sharp edge for the purpose of shaving and generating a material chip (curl). This edge must remain sharp for the tool to perform properly. Grinding tools are different in that they have an abrasive-coated surface which generates a powder (swarf) as opposed to the chip of a cutting tool. Diamond performs well in these two basic operations but with limitations as shown below.[17]

The three requirements of a cutting or grinding tool material are hardness, toughness, and chemical stability. Diamond meets the first since it is the hardest material. However, it is inherently brittle, has low toughness, and reacts readily with carbide-forming metals, thus limiting its use.

Diamond Toughness and Bonding Techniques. Several techniques are available to bond diamond to a substrate such as resinous, vitrified, and electroplated-metal bondings. These techniques impart different degrees of toughness to the bond as shown Fig. 12.9, and the selection of the proper bond is essential to achieve optimum results in machining or grinding workpieces of different moduli of *resilience*.[17][18]

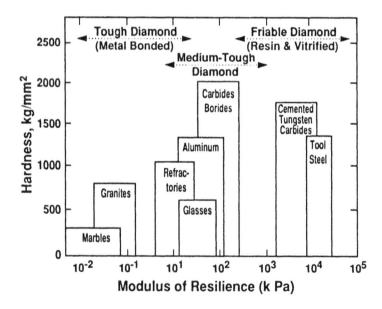

Figure 12.9. Selection of diamond type as a function of modulus of resilience of workpiece.[18]

The machining characteristics of diamond are also influenced by the shape and size of the crystal, and manufacturers offer a great variety of crystals specifically suited to grinding, cutting, sawing, drilling, or polishing.[19] The selection of grain size for various applications is shown in Fig. 12.10.[20] The cutting material employed in the new ultraprecision machining technique of single-point turning is single-crystal diamond.[21]

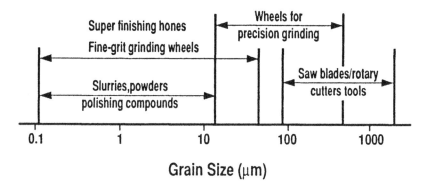

Grain Size (μm)

Figure 12.10. Selection of diamond-abrasive grain size for various applications.[20]

Chemical Stability of Diamond Tools. Diamond reacts with carbide-forming metals such as iron. In contact with these, diamond dissolves rapidly and is considered unsuitable to machine steel and cast iron. Likewise, it is not recommended for superalloys.

Polycrystalline Diamond. Polycrystalline diamond (PCD) is an aggregate produced under heat and pressure from single crystals varying in size from 5 to 20 μm, bonded together with cobalt or nickel.[22] Unlike the brittle single crystals, PCD is tough and well-suited for demanding applications. It is similar to the natural *carbonado* described in Sec. 2.3.

PCD can be produced as blanks, cut into specific shapes, and bonded to a tungsten carbide substrate for a fast-growing range of applications such as machining and cutting ceramics and ceramic composites, and oil-drilling bits (Fig. 12.11).

Figure 12.11. Polycrystalline-diamond drilling products "Stratapax." *(Photograph courtesy of GE Superabrasives, Worthington, OH)*

Applications of Diamond Tools. Most industrial machining opera-tions are performed with coated cemented-carbide tools, usually as indexable inserts. Diamond (and PCD) increasingly competes with these materials, in particular in the machining of non-ferrous metals such as aluminum, copper, magnesium, lead, and their alloys since it does not react chemically with these materials.

Of particular importance is the machining of aluminum alloys now widely used in automobiles and other areas. These alloys are difficult to machine with carbides, especially if a good finish is required, and diamond has shown to be the best machining material. Diamond is used extensively in machining abrasive materials, such as fiber composites, ceramics, cemented carbides, graphite, concrete, and silicon. A major application is the cutting of architectural stones.

Diamond tools, although the best performers in many instances, are expensive and require excellent machine conditions, such as closely controlled speed and the absence of chatter and vibration, for optimum performance.

5.3   Thermal Management (Heat Sink) Applications

Because of its high thermal conductivity, diamond is an excellent heat-sink material, mostly for high-density electronic circuits. A thin slice, in size up to 3 x 3 x 1 mm, is cut from a natural crystal, metallized by sputtering a gold alloy and soldered to a copper base. A typical diamond heat-sink configuration is shown in Fig. 12.12.[23]

The drawbacks of these single-crystal diamond heat sinks are the presence of metallic impurities and size limitations. CVD diamond, which does not have these drawbacks, is a competitor which may eventually replace the single-crystal material (see Ch. 13).

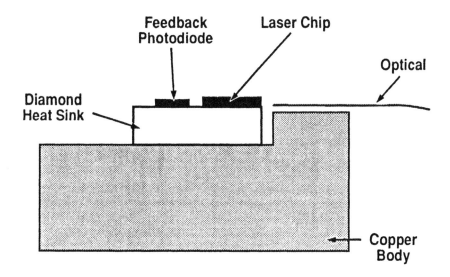

Figure 12.12. Typical design of laser diode on natural-diamond heat-sink.[23]

5.4   Miscellaneous Applications

Natural diamond is used almost exclusively for applications requiring half a carat or more such as heat sinks, optical windows, knives, and others described below. Such applications form a small but steadily growing market although, here again, the possibility of being replaced by CVD diamond should not be overlooked.

Optical Applications. Although diamond can be considered as the ideal optical material, it has found little applications in optics so far because of its restricted size and high cost. It has been used by NASA as an optical window for the Pioneer Venus spacecraft, clearly a case where cost is secondary to performance.[24]

Mechanical Applications. Anvils made of single-crystal natural diamond can generate extremely high pressures, up to 4.6 Mbar (a pressure significantly higher than the pressure at the center of the earth). Typical anvil shapes are shown in Fig. 12.13.[24]

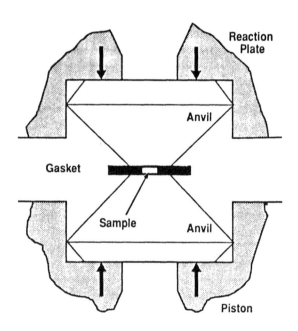

Figure 12.13. Schematic of a natural-diamond anvil and high-pressure cell.[24]

Another mechanical application is found in surgical knives and scalpels. In the process of cutting, the object to be cut is stretched by the edge of the blade until it ruptures in tension. The sharper the edge, the less force is required to cut and the more accurate the cut is. Diamond, with its strong interatomic bond, provides edges which are up to one hundred times sharper than the finest steel edge. This is very important in surgery, particularly eye and nerve surgery, where diamond is replacing steel although its cost is much higher. In addition, diamond is transparent and can be used to illuminate the area of the cut and transmit laser energy for the treatment of tumors and other diseases. A schematic of an experimental blade is shown in Fig. 12.14.[24]

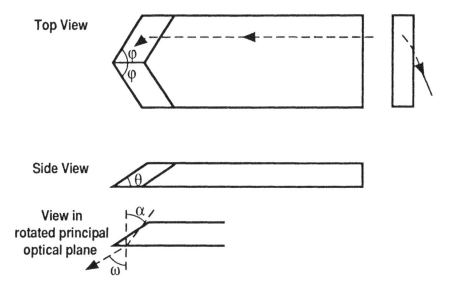

Figure 12.14. Schematic of an experimental surgical natural-diamond blade showing laser-light path.[24]

REFERENCES

1. Herbert, I., *The Diamond Diggers*, Tom Stacey Pub., London (1972)

2. Davies, G., *Diamond*, Adams Hilger Ltd., Bristol, UK (1984)

3. Krauskopf, K. B., *Introduction to Geochemistry*, McGraw-Hill, New York (1967)

4. Dawson, J. B., *The Properties of Diamond* (J. E. Field, ed.), 539-554, Academic Press, London (1979)

5. Orlov, Y. L., *The Mineralogy of Diamond*, John Wiley & Sons, New York (1977)

6. Austin, G. T., *Ceramic Bull.*, 70(5):860 (1991)

7. Bundy, F. P., Strong, H. M. and Wentorf, R. H., Jr., *Chemistry and Physics of Carbon*, Vol. 10 (P. L., Walker, Jr. and P. A. Thrower, eds.), Marcel Dekker Inc. (1973)

8. Spear, K. E., *J. Amer. Ceram. Soc.*, 72(2):171-191 (1989)

9. Adams, D. M., *Inorganic Solids*, John Wiley & Sons, New York (1981)

10. Hirano, S. and Naka, S., *Fine Ceramics* (S. Saito, ed.), 2-15, Elsevier-Ohmsha, New York (1988)

11. Muncke, G., *The Properties of Diamond*, (J. E. Field, ed.), 473-499, Academic Press, London (1979)

12. Wedlake, R. J., *The Properties of Diamond*, (J. E. Field, ed.), 501-535, Academic Press, London (1979)

13. Cowan, G., Dunnington, B., and Holtzman, A., "Process for Synthesizing Diamond," U.S. Patent 3,401,019 (Sept. 10, 1968)

14. Busch, J. V., Dismukes, J. P., Nallicheri, N. V., and Walton, K. R., *Applications of Diamond Films and Related Materials*, (Y. Tzeng, et al., eds.), 623-633, Elsevier Science Publishers (1991)

15. Seal, M., *Applications of Diamond Films and Related Materials*, (Y. Tzeng, et al., eds.), 3-7, Elsevier Science Publishers (1991)

16. Fritsch, E., Shigley, J. E., and Koivula, J. I., *Diamond Optics*, 969:114-116, SPIE (1988)

17. Peterman, L., *Proc. SPI/RP-Ci, 41st. Annual Conf.* (Jan 1986)

18. Gardinier, C. F., *Ceramic Bulletin*, 67(6):1006-1009 (1988)

19. McEachron, R. W. and Lorence, S. C., *Ceramic Bulletin*, 67(6):1031-1036 (1988)

20. Subramanian, K., *Ceramic Bulletin* 67(6):1026-1029 (1988)

21. Ikawa, N., Shimada, S., and Morooka, H., T, *Bull. Japan Soc. of Prec. Eng.*, 21(4):233-238 (1987)

22. Schaible, J., *Cutting Tool Engineering*, 41(3):106-109 (June 1989)

23. *Diamond Heat Sinks,* Technical Brochure from Drukker & Zn. NV, 1001 MC. Amsterdam, The Netherland (1989)

24. Seal, M., *Interdisciplinary Science Review*, 14(1):64-76 (1989)

13

CVD Diamond

1.0 INTRODUCTION

The production of diamond coatings is now possible by low-pressure vapor-phase synthesis under relatively benign conditions. This development should open the door to the full exploitation of the unique properties of diamond at a cost that is still high but that will likely be reduced as technology improves and production is achieved on an industrial scale. At this time, the technology remains mostly at the laboratory stage and only a few applications are on the market. Yet the future appears promising with a broad range of potential applications in many industrial areas such as heat sinks for electronic devices, advanced semiconductors and tool coatings.

The low-pressure vapor-phase process is based on chemical vapor deposition (CVD) and the material is often referred to as "vapor-phase diamond," "diamond coating," or "CVD diamond". "CVD diamond" will be used in this book.

Diamond coatings are also produced by physical vapor deposition (PVD). Such coatings however are a mixture of sp^2 (graphite) and sp^3 (diamond) bonds and are considered a different material usually referred to as diamond-like carbon (DLC). They are reviewed in Ch. 14.

CVD-diamond coatings are polycrystalline, as opposed to natural and high-pressure synthetic diamond which are normally single crystals. This polycrystalline characteristic has important bearing on the general properties of the coatings as shown in Sec. 4.0.

1.1 Historical Perspective

As seen in Ch. 12, the high-pressure synthesis of diamond became an industrial reality in 1957 when General Electric started production. It was generally believed at that time in the U.S. that low-pressure synthesis was not feasible or at least not practical and early inconclusive work at Union Carbide and Case Western University were not actively followed.[1][2]

In the former Soviet Union, a less skeptical attitude prevailed and a research program was aggressively pursued, particularly in the investigation of hydrocarbon-hydrogen pyrolysis and nucleation and growth mechanism of diamond. A milestone was the discovery by Soviet researchers of the role of atomic hydrogen in removing unwanted graphite but leaving diamond untouched (see Sec. 2.2). The need to increase the availability of atomic hydrogen soon became apparent and the Soviets developed various methods to achieve that end, such as electric discharge and heated filaments.[3]-[5]

The Soviet successes were largely ignored in the U.S. but not so in Japan where they triggered considerable interest. The Japanese, under the leadership of the National Institute for Research in Inorganic Materials (NIRIM), started a large, well-coordinated scientific effort sponsored by government agencies, universities, and industry. Japan is still a leader in this technology, and research and development in the former Soviet Union is also continuing on an important scale.

In the U.S. interest is now high and the technology is developing at a fast rate, with leadership found in universities (particularly Pennsylvania State and North Carolina State) and in many industrial organizations.[6]

Some of the major historical developments of CVD diamond are summarized in Table 13.1.

1.2 CVD-Diamond Coatings

At the present, CVD diamond is produced in the form of coatings. In this respect, it is similar to pyrolytic graphite (see Ch. 7, Sec. 1.3). These coatings can alter the surface properties of a given substrate as shown in Table 13.2. This table may be compared to Table 7.1 of Ch. 7 in which the critical properties of pyrolytic graphite coatings are listed.

Coatings with a thickness of 1 mm or more are now routinely produced. After removing the substrate, a free-standing shape remains with good integrity and properties similar to single-crystal diamond as reviewed in Sec. 5.6.

Table 13.1. Chronology of Major Developments of CVD Diamond

1950's	Early work on low-pressure synthesis mostly at Union Carbide, Case Western, and in the Soviet Union
1955	First production of high-pressure synthetic diamond in Sweden, the U.S. and the Soviet Union
1956-present	Continuing development of low pressure synthesis by Derjaguin and others in the Soviet Union
1974-present	NIRIM and other Japanese laboratories develop high growth rate by CVD process
1985	Consortium formed at Pennsylvania State University to promote diamond research in the U.S.
1988	Development of diamond tweeter diaphragm by Sumitomo Electric
1989	Development of diamond-coated boring and drilling tools by Mitsubishi Metals
1992	Commercial production of free-standing shapes up to 1 mm thick by Norton, General Electric, and others

Table 13.2. Material Properties Affected by CVD-Diamond Coatings

Electrical	Resistivity Dielectric constant
Optical	Refraction Emissivity Reflectivity Selective absorption
Mechanical	Wear Friction Hardness Adhesion Toughness Strength
Chemical	Corrosion resistance

2.0   DEPOSITION MECHANISM OF CVD DIAMOND

The principles of thermodynamics and kinetics of CVD as they apply to the deposition of graphite were reviewed in Ch. 7, Sec. 2.0.[7]   These considerations are valid for the CVD of diamond as well.

2.1   Basic Reaction

The basic reaction in the CVD of diamond is seemingly simple.   It involves the decomposition of a hydrocarbon such as methane as follows:

$$CH_4 \text{ (g)} \xrightarrow[\text{activation}]{} C \text{ (diamond)} + 2H_2 \text{ (g)}$$

In reality, the deposition mechanism is a complex interaction of many factors and is not fully understood at this time, although two essential conditions have been identified: *(a)* activation of the carbon species and *(b)* the action of atomic hydrogen.   These factors are reviewed in the following sections.

2.2   Deposition Mechanism and Model

Diamond has been deposited from a large variety of precursors which include, besides methane, aliphatic and aromatic hydrocarbons, alcohols, ketones, and solid polymers such as polyethylene, polypropylene and polystyrene (a special case, deposition with halogens, is reviewed in Sec. 2.5).

These substances generally decompose into two stable primary species: the methyl radical (CH_3) and acetylene (C_2H_2).[6]   The methyl radical is considered the dominant species and the key compound in generating the growth of CVD diamond, at least in the hot-filament deposition process (see Sec. 3.4).[8][9]   Direct deposition from acetylene, although difficult experimentally, has been accomplished, with a marked increase in the crystallinity of the diamond deposit.[10]

Deposition Model.   A two-step deposition model was recently presented by Frenklash and Spear.[11]   In the first step, the diamond surface is activated by the removal of a surface-bonded hydrogen ion by atomic hydrogen as follows:

$$\text{H.} + \begin{array}{c} \text{C} \\ \diagdown \\ \text{C-H} \\ \diagup \\ \text{C} \end{array} \rightarrow \text{H}_2 + \begin{array}{c} \text{C} \\ \diagdown \\ \text{C.} \\ \diagup \\ \text{C} \end{array}$$

In the second step, the activated surface-carbon radical reacts with the carbon-hydrogen species (acetylene as a monomer unit) in the gas phase to become the site for carbon addition:

$$\begin{array}{c} \text{C} \\ \diagdown \\ \text{C.} \\ \diagup \\ \text{C} \end{array} + \text{C}_2\text{H}_2 \rightarrow \begin{array}{c} \text{C} \quad\quad \text{H} \quad \text{H} \\ \diagdown \quad\quad | \quad\quad / \\ \text{C - C = C.} \\ \diagup \\ \text{C} \end{array}$$

The model is consistent with experimental observations and should provide a useful guideline for future experiments. A similar model has been proposed that is based on the addition of a methyl group to one of the carbons followed by atomic hydrogen abstraction from the methyl group.[12][13]

2.3   Role of Atomic Hydrogen

Hydrogen, in normal conditions, is a diatomic molecule (H_2) which dissociates at high temperature (i.e., >2000°C) or in a high current-density arc to form atomic hydrogen. The dissociation reaction is highly endothermic (ΔH = 434.1 kJmol^{-1}).

The rate of dissociation is a function of temperature, increasing rapidly above 2000°C. It also increases with decreasing pressure.[10] The rate of recombination (i.e., the formation of the molecule) is rapid since the mean-free-path dependent half-life of atomic hydrogen is only 0.3 s.

As shown in the models reviewed above, atomic hydrogen plays an essential role in the surface and plasma chemistry of diamond deposition as it contributes to the stabilization of the sp^3 dangling bonds found on the diamond surface plane (Fig. 13.1).[4] Without this stabilizing effect, these bonds would not be maintained and the diamond {111} plane would collapse (flatten out) to the graphite structure.

The other function of atomic hydrogen is to remove graphite selectively. In contrast with molecular hydrogen, atomic hydrogen is extremely reactive. It etches graphite twenty times as fast as it etches diamond and even faster in the presence of oxygen. This etching ability is important since, when graphite and diamond are deposited simultaneously, graphite is preferentially removed while most of the diamond remains.[15]

These two effects of atomic hydrogen, graphite removal and sp³ bond stabilization, are believed essential to the growth of CVD diamond.

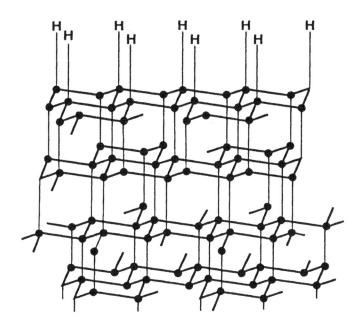

Figure 13.1. Schematic of CVD diamond structure showing hydrogen atoms bonded to growth surface.

2.4   Effect of Oxygen and Oxygen Compounds

Beside the need for atomic hydrogen, the presence of oxygen or an oxygen compound such as H_2O, CO, methanol, ethanol, or acetone, can be an important contributor to diamond film formation. A small amount of oxygen added to methane and hydrogen tends to suppress the deposition of graphite by reducing the acetylene concentration as well as increasing the diamond growth rate.[6][16] The addition of water to hydrogen appears to increase the formation of atomic hydrogen which would explain the observed increased deposition rate.

2.5   Halogen-Based Deposition

It was recently found that diamond growth also occurs in several halogen-based reactions. These reactions proceed at lower temperature (250 - 750°C) than those based on the methyl-radical mechanism reviewed above. The halogen reaction mechanism is still controversial and the optimum precursor species are yet to be determined.[17]

To proceed, the reactions must be highly favored thermodynamically. This is achieved when the reaction products are solid carbon and stable gaseous fluorides or chlorides (HF, HCl, SF_6). Typical reactions and their free energy at 1000 K are shown in Table 13.3. These free-energy values are for the formation of graphite and not diamond. However the diamond free-energy of formation is not much higher than that of graphite ($\Delta G = 2.87$ kJ/mol).

The deposition reaction is carried out in a simple flow tube. The amount of carbon-containing gas is maintained at <5% of the overall gas composition to retard formation of non-diamond carbon. The addition of oxygen or oxygen compounds (air, H_2O, CO_2) enhances growth.

Diamond growth with the halogen reactions has also been observed with the hot-filament, RF-discharge, and microwave processes.[18]-[20] These processes are described in the following section.

Table 13.3.   Diamond-Forming Halogen Reactions and Their Free-Energy[17]

Reaction	$\Delta G°$ at 1000 K (kJ/mol)
$CH_4 \rightarrow C + 2H_2$ (for comparison)	-4.5
$CH_4 + 2F_2 \rightarrow C + 4HF$	-1126.5
$CCl_2F_2 + 2H_2 \rightarrow C + 2HCl + 2HF$	-407.6
$CH_3OH + F_2 \rightarrow C + H_2O + 2HF$	-725.9
$CH_3CH_2OH + 2F_2 \rightarrow 2C + H_2O + 4HF$	-1358.7
$CH_3SH + 5F_2 \rightarrow C + 4HF + SF_6$	-2614.4
$CS_2 + 6F_2 \rightarrow C + 2SF_6$	-3407.2

3.0 CVD DIAMOND PROCESSES

3.1 General Characteristics

As mentioned above, the carbon species must be activated since at low pressure, graphite is thermodynamically stable, and without activation, only graphite would be formed (see Ch. 11, Sec. 3.2). Activation is obtained by two basic methods: high temperature and plasma, both of which requiring a great deal of energy.

Several CVD processes based on these two methods are presently in use. These processes are continuously being expanded and improved and new ones are regularly proposed. The four most important at this time are: high-frequency (glow) plasma, plasma arc, thermal CVD, and combustion synthesis (oxy-acetylene torch). Their major characteristics are summarized in Table 13.4.[6]

Table 13.4. Characteristics of Diamond Deposition Processes

Activation Method	Process	Substrate Deposition Rate	Temperature Control	Main Product
Glow-discharge plasma	Microwave RF	Low (0.1 - 10 μm/h)	Good	Coating
Arc plasma	DC Arc RF Arc	High (50 - 1000 μm/h)	Poor Plates	Coating
Thermal	Hot-filament	Low (0.1 - 10 μm/h)	Good Plates	Coating
Combustion	Torch	High	Poor	Coating Powder

3.2   Types of Plasma

Most diamond-deposition processes require a plasma. As the temperature of a gas is increased, its atoms are gradually ionized, that is, they are stripped of their electrons and a plasma is formed which consists of ions (positive charge), electrons (negative charge), and atoms that have not been ionized (neutral).[21]

Two types of plasma are currently used for the deposition of diamond: glow-discharge plasma (non-isothermal) and arc plasma (isothermal). Their characteristics are shown in Table 13.5 and are described in more detail in the following sections.

Table 13.5. Characteristics of Plasmas for Diamond Deposition[21]

	Glow-Discharge	Arc
Plasma Type	Non-Isothermal (non-equilibrium)	Isothermal (equilibrium)
Frequency	50 kHz - 3.45 MHz and 2.45 GHz (microwave)	~ 1 MHz
Power	1 - 100 kW	1 - 20 MW
Flow rate	mg/s	none
Electron concentration	10^9 - 10^{12}/cm^3	10^{14}/cm^3
Pressure	200 Pa - 0.15 atm	0.15 - 1 atm
Electron temperature	10^4 K	10^4 K
Atom temperature	500 K	10^4 K

3.3 Glow-Discharge (Microwave) Plasma Deposition

A glow-discharge (non-isothermal) plasma is generated in a gas by a high-frequency electric field such as microwave at relatively low pressure. In such a plasma, the following events occur:

- In the high-frequency electric field, the gases are ionized into electrons and ions. The electrons, with their very small mass, are quickly accelerated to high energy levels corresponding to 5000 K or higher.

- The heavier ions with their greater inertia cannot respond to the rapid changes in field direction. As a result, their temperature and that of the plasma remain low, as opposed to the electron temperature (hence the name non-isothermal plasma).

- The high-energy electrons collide with the gas molecules with resulting dissociation and generation of reactive chemical species and the initiation of the chemical reaction.

The most common frequencies in diamond deposition are the microwave (MW) frequency at 2.45 GHz and, to a lesser degree, radio frequency (RF) at 13.45 MHz (the use of these frequencies must comply with federal regulations).

Deposition Process. A typical microwave plasma for diamond deposition has an electron density of approximately 10^{20} electrons/m^3, and sufficient energy to dissociate hydrogen. A microwave-deposition reactor is shown schematically in Fig. 13.2.[16][22] The substrate (typically a silicon wafer) is positioned at the lower end of the plasma. Gases are introduced at the top of the reactor, flow around and react at the substrate, and the gaseous by-products are removed into the exhaust. The substrate must be heated to 800 - 1000°C for diamond to form. This can be done by the interaction with the plasma and microwave power but this is difficult to regulate and, more commonly, the substrate is heated directly by radiant or resistance heaters which provide more accurate temperature control.

Typical microwave deposition conditions are the following:

Incident Power:	600 W
Substrate Temp.:	800 - 1000°C
Gas mixture H_2/CH_4:	50/1 to 200/1
Pressure:	10 to 5000 Pa
Total gas flow:	20 - 200 scm^3/min

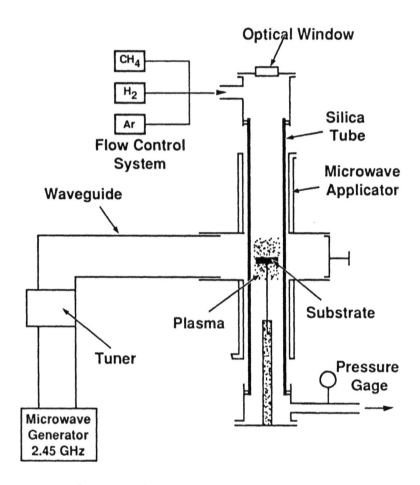

Figure 13.2. Schematic of microwave-plasma deposition apparatus.

The morphology and properties of the deposited coating vary as a function of substrate temperature, gas ratio, and the intensity of the plasma at the deposition surface. Deposition rate is low, averaging 1 - 3 μm/h. This may be due to the limited amount of atomic hydrogen available in the deposition zone (estimated at 5%).

An advantage of microwave plasma is its stability which allows uninterrupted deposition lasting for days if necessary. However the plasma can be easily disturbed by the addition of oxygenated compounds.

Electron Cyclotron Resonance (ECR). A microwave plasma can also be produced by electron cyclotron resonance (ECR), through the proper combination of electric and magnetic fields. Cyclotron resonance is achieved when the frequency of the alternating electric field is made to match the natural frequency of the electrons orbiting the lines of force of the magnetic field. This occurs at the standard microwave frequency of 2.45 GHz with a magnetic field of 875 Gauss. An ECR plasma reactor suitable for the deposition of diamond is shown schematically in Fig. 13.3.[23]

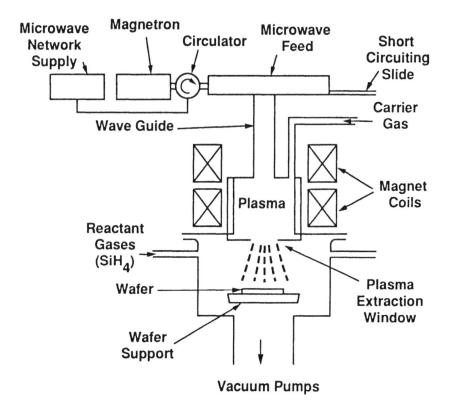

Figure 13.3. Schematic of electron-cyclotron-resonance (ECR) apparatus for the deposition of diamond.[23]

An ECR plasma has two basic advantages for the deposition of diamond: (a) it minimizes the potential substrate damage caused by high-intensity ion bombardment, usually found in an standard high-frequency plasma where the ion energy may reach 100 eV, (b) it minimizes the risk of damaging heat-sensitive substrates since it operates at a relatively low temperature.

Disadvantages are a more difficult process control and more costly equipment due to the added complication of the magnetic field.

RF Plasma. The activation of the reaction and the generation of atomic hydrogen can also be obtained with an RF plasma (13.56 MHz) but this process is more likely to produce diamond-like carbon (DLC) and not pure diamond (see Ch. 14).

3.4 Plasma-Arc Deposition

In addition to microwave deposition, another common plasma deposition system for diamond coatings is based on plasma-arc. Plasma-arc deposition is usually obtained in a high-intensity, low-frequency arc, generated between two electrodes by either direct or alternating current. The process requires a large amount of power and the equipment is costly.[24]

In a low-frequency plasma, both electrons and ions respond to the constantly, but relatively slowly, changing field direction, as opposed to a high-frequency plasma where only the electrons respond. Both electrons and ions acquire energy and their temperature is raised more or less equally. The plasma is in equilibrium (isothermal) as opposed to the non-isothermal condition found in a high-frequency plasma. Isothermal plasmas for diamond deposition are generated at a higher pressure than glow-discharge plasmas (0.15 to 1 atm). At such pressure, the average distance traveled by the species between collisions (mean free path) is reduced and, as a result, molecules and ions collide more frequently and heat more readily.

By increasing the electrical energy in a fixed amount of gas, the temperature is raised and may reach 5000°C or higher.[21] Such high temperatures produce an almost complete dissociation of the hydrogen molecules, the CH radicals, and other active carbon species. From this standpoint, arc-plasma systems have an advantage over glow-discharge or thermal CVD since these produce a far smaller ratio of atomic hydrogen.

DC Plasma System. Typical direct-current (DC) plasma deposition systems are shown schematically in Fig. 13.4.[25][26] Electrodes usually consist of a water-cooled copper anode and a tungsten cathode. Several gas-jet nozzles can be operated simultaneously and many design variations are possible, including separate input nozzles for hydrogen and methane (the latter mixed with argon) and the feeding of these gases in a coaxial feed electrode.

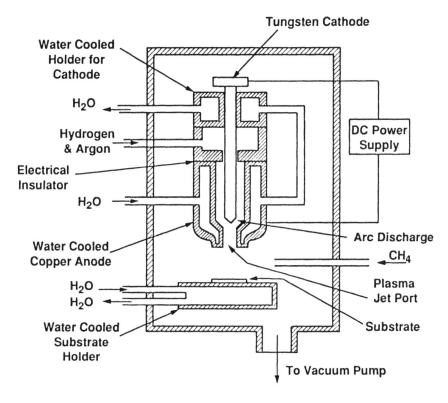

Figure 13.4. Schematic of arc-discharge apparatus for the deposition of diamond.[26]

Another system incorporates the interaction of a solenoid magnetic field to give the arc a helical shape. This stabilizes and increases the length of the arc.[27]

The plasma jet can be cooled rapidly just prior to coming in contact with the substrate by using a blast of cold inert gas fed into an annular fixture. Gaseous boron or phosphorus compounds can be introduced into the gas feed for the deposition of doped-semiconductor diamond.[28]

The sudden expansion of the gases as they are heated in the arc plasma causes the formation of a high-speed arc jet so that the atomic hydrogen and the reactive carbon species are transported almost instantly to the deposition surface and the chances of hydrogen recombination and of vapor-phase reactions are minimized.

The substrate may be heated to unacceptable levels by the high temperature of the gases and cooling is usually necessary. Temperature control and substrate cooling remain a problem in arc-plasma systems. However, deposition is rapid and efficient, high rates of deposition (80 μm/ h or higher) are possible, and thick deposits are routinely produced.[29] The availability of free-standing shapes, 15 cm in diameter and 1 mm thick, has recently been announced.[30][31] Typical shapes are shown in Fig. 13.5.

Figure 13.5. Typical free-standing CVD diamond shapes. *(Photograph courtesy GE Superabrasives, Worthington, OH.)*

3.5 Thermal CVD (Hot Filament)

In the processes described above, the plasma is generated by an electric current. A plasma can also be generated by high temperature which, in the case of diamond deposition, is obtained by a resistively heated wire or tube made of tungsten or tantalum.[17][32][33] A schematic of the equipment is shown in Fig. 13.6.

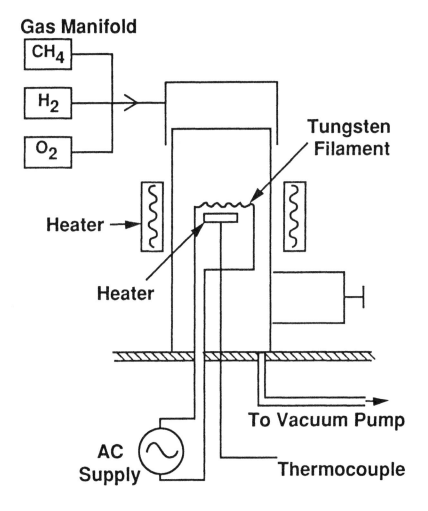

Figure 13.6. Schematic of hot-filament apparatus for the deposition of diamond.[32][33]

The metal temperature is maintained at 2000°C or slightly higher. Atomic hydrogen is formed and the carbon species become activated in the vicinity of the hot metal. The deposition rate and the composition and morphology of the deposit are functions of the temperature and the distance between the hot metal and the substrate. This distance is usually 1 cm or less. Much beyond that, most of the atomic hydrogen recombines and no diamond is formed.

The substrate temperature should be kept between 800 and 1000°C and cooling may be necessary. Gas composition and other deposition parameters are similar to those used in a microwave-plasma system. Deposition rate is low, reported as 0.5 to 1 μm/h. A disadvantage of the hot-filament process is the short life of the metallic heater, which tends to carburize, distort, and embrittle rapidly. In this respect, tantalum performs better than tungsten with an estimated life of 600 hrs (vs. 100 hrs for tungsten).[33] The heated metal may also evaporate and contaminate the diamond film. Furthermore, it is not advisable to add oxygen or an oxygen compound as mentioned in Sec. 2.4 since, at these temperatures, tungsten (or most other refractory metals) would oxidize rapidly. However, the equipment is relatively inexpensive and experiments are readily carried out. Other heating-element materials such as graphite or rhenium are being investigated.[34]

3.6 Combustion Synthesis (Oxy-Acetylene Torch)

Diamond is grown in air with a simple unmodified oxy-acetylene brazing/welding torch.[35][36] Substrates such as silicon can be rapidly coated when exposed to the reducing portion of the flame but uniformity in structure and composition is not readily achieved.

The high gas temperature (>2000°C) makes it mandatory to cool the substrate. As a result, large thermal gradients are produced which are difficult to control. The deposition efficiency is extremely low with a nucleation rate of $1/10^6$. This means high gas consumption, high energy requirements, and high cost. The deposition mechanism is not clearly understood at this time.

3.7 Diamond from ^{12}C Isotope

CVD diamond has been produced from the single carbon isotope, ^{12}C. As a reminder, normal diamond is composed of 98.89% ^{12}C and 1.108% ^{13}C (see Ch. 2, Sec. 2.1). The deposition of ^{12}C diamond is accomplished by the decomposition in a microwave plasma of methane in which the carbon atom is 99.97% ^{12}C. The resulting polycrystalline deposit is crushed, processed to a single crystal with molten iron and aluminum at high temperature and pressure, and recovered by leaching out the metals[37][38] (See Ch. 12, Sec. 3.0).

In diamond, thermal conductivity occurs by a flow of phonons (see Ch. 11, Sec. 5.2). These phonons are scattered by imperfections such as isotopic imperfections and the scattering varies as the fourth power of the phonon frequency. Thus, the exclusion of ^{13}C should result in a significant increase in thermal conductivity. This increase was confirmed experimentally as ^{12}C diamond is reported to have a 50% higher thermal conductivity than natural diamond.[37][38] It is also reported to be harder than natural diamond by a few percent as determined by the relation between hardness and the elastic coefficients (see Ch. 11, Sec. 10.2).

3.8 Nucleation and Structure

CVD diamond is a polycrystalline material with grain size ranging from a few nanometers to a few micrometers. The polycrystallinity results from the formation of separate nuclei on the deposition surface. As deposition proceeds, these nuclei eventually coalesce to form a continuous film. Some grow in size while other are crowded out and the average crystal size often becomes larger as thickness increases. Some degree of crystal orientation (texture) can be achieved. A cross-section of a coating is shown in Fig. 13.7.

As mentioned in Ch. 11, the most prevalent crystal surfaces in CVD diamond are the (111) octahedral and the (100) cubic. Cubo-octahedral crystals combining both of these surfaces are also common. Twinning occurs frequently at the (111) surface. The 6H structure may also be present.

3.9 Substrate Preparation and Adhesion

The nature of the substrate and its predeposition treatment play a major role in determining the surface nucleation rate but not necessarily the rate of subsequent growth (after the immediate surface layer is deposited).[13][39] The most widely used substrate is still silicon, but other materials perform successfully such as the refractory metals (W, Ta, Mo), carbides

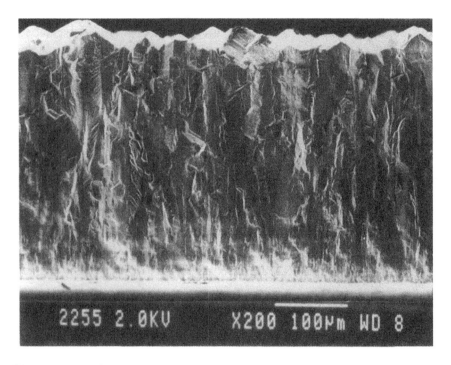

Figure 13.7.   Cross-section of diamond coating.   *(Photograph courtesy of Diamonex, Allentown, PA.)*

(WC, SiC), and other metals such as Cu, Au and Ni and their alloys.[40]   Of particular industrial importance is the deposition on tool-steel substrate (see Sec. 5.0).   Cubic boron nitride (c-BN) is also a suitable substrate with good lattice matching but is only available as small single crystals.

The nucleation rate and the adhesion vary with the nature of the substrate and appear to be related to the ability of the substrate material to form an intermediate carbide.   Surface treatments such as etching or mechanical working (scratching with a diamond powder or diamond polish) help promote adhesion.

3.10 Oriented and Epitaxial Growth

The properties of diamond are affected by impurities and lattice defects.   In addition, they are influenced by crystal boundaries, especially if high-angle boundaries are dominant. A polycrystalline material such as CVD diamond has many such boundaries and various crystallite orientations and grain sizes. As a result, its properties (particularly thermal, optical, and electrical) do not quite match those of single-crystal diamond, as shown in Table 13.6.

The ability to deposit single-crystal diamond, or at least a material with a high degree of crystalline orientation and with properties equal to the high-purity single crystal material, would be an important factor in the development of CVD diamond in electronic, semiconductor, optical, and other applications.

The epitaxial growth of single-crystal diamond by CVD could be accomplished by selecting a substrate with a crystal lattice parameters that closely matches that of diamond and that would have the same atomic bonding characteristic.[28] Single-crystal diamond is the only substrate found suitable so far but it is limited by its small size and high cost. The development of other suitable substrates for epitaxial growth would allow the production of diamond semiconductor wafers. Indications are that some nickel alloys such as nickel-copper have a crystal structure that is sufficiently close to that of diamond to allow the deposition of epitaxial films.[41]

3.11 Morphology

The as-deposited surface of CVD diamond is rough as shown in Fig. 13.8 and a polishing step is required for many applications. Polishing to less than 5 nm radius without chipping or grain pulling is normally a difficult task. A smooth surface is now possible with the development of an ion-implantation technique with an energy range of up to 5000 keV, which softens the surface. This step is followed by a mechanical polishing with diamond paste.[42]

4.0 PROPERTIES OF CVD DIAMOND

4.1 Summary of Properties

CVD diamond is a recent material and, like single-crystal diamond, it is expensive and available only in small quantities and mostly in the form of coatings. As a result, the evaluation of its properties is still limited. The

Figure 13.8. Scanning electron micrograph of CVD-diamond coating. *(Photograph courtesy Norton Diamond Film, Northboro, MA.)*

difficulty in testing, the effect of impurities and structural defects, and the differences between the various deposition processes may contribute to the uncertainty and spread found in the reported property values.

These properties are summarized and compared to single-crystal diamond in Table 13.6.[30]-[33][43]

4.2 Thermal Properties

The mechanism of thermal conductivity in CVD diamond was reviewed in Secs. 3.6 and 3.9. In spite of crystal boundaries and resulting phonon scattering, the thermal conductivity is remarkably high which makes CVD diamond particularly suitable for heat-sink applications (see Sec. 5.4).

Table 13.6. Properties of CVD diamond

	CVD Diamond	Single-Crystal Diamond
Density, g/cm^3	3.51	3.515
Thermal conductivity at 25°C, W/m·K	2100	2200
Thermal expansion coefficient x 10^{-6}/°C @ 25 - 200°C	2.0	1.5 - 4.8
Bangap, eV	5.45	5.45
Index of refraction at 10 μm	2.34 - 2.42	2.40
Electrical resistivity, ohm·cm	10^{12} - 10^{16}	10^{16}
Dielectric constant (45 MHz - 20 GHz)	5.6	5.70
Dielectric strength, V/cm	10^6	10^6
Loss tangent (45 MHz - 20 GHz)	<0.0001	
Saturated electron velocity	2.7	2.7
Carrier mobilities (cm^2/V·s) electron (n) positive hole (p)	1350 - 1500 480	2200 1600
Vickers hardness range*, kg/mm^2	5000 - 10000	5700 - 10400
Coefficient of friction	0.05 - 0.15	0.05 - 0.15

* Varies with crystal orientation

4.3   Optical Properties

In many optical, electronic, and opto-electronic applications, the main emphasis is not so much on thickness since the required thickness is submicron or at the most a few μm, but more on the deposition of pure diamond crystals without carbon, graphite, or metallic impurities. This has yet to be achieved and present materials are opaque or translucent and certainly not on par with gem-quality material[39] (see Ch. 11, Sec. 6.0).

4.4 Electronic and Semiconductor Properties

The semiconductor characteristics of CVD diamond are similar to those of the single crystal (see Ch. 11, Sec. 9.3). It is an *indirect-bandgap, high-temperature* semiconductor which can be changed from an intrinsic to an extrinsic semiconductor at room temperature by doping with other elements such as boron or phosphorus.[6][28] This doping is accomplished during deposition by introducing diborane (B_2H_6) or phosphine (PH_3) in the deposition chamber. The semiconductor properties of CVD diamond are similar to those of the single crystal.

4.5 Mechanical Properties

The hardness of single-crystal diamond varies as a function of the crystal orientation by almost a factor of two. This is also true of CVD diamond, and the hardness values depend on which crystal face is in contact with the indenter of the testing device.

Wear resistance of CVD diamond is generally superior to that of the single-crystal material since the wear of diamond occurs by chipping. Since CVD diamond is a polycrystalline material, chipping stops at the grain boundary while, in a single-crystal, the entire crystal is sheared off.

4.6 Chemical Properties

The chemical properties of CVD diamond are similar to those of the single-crystal material reviewed in Ch. 11, Sec. 11.0.

5.0 APPLICATIONS OF CVD DIAMOND

5.1 Status of CVD Diamond Applications

The applications of natural and high-pressure synthetic diamond were reviewed in Ch. 12. Although these applications have a very large market, particularly in gemstones, they are limited because of the small size and high cost of the crystals. CVD diamond, on the other hand, offers a broader potential since size, and eventually cost, are less of a limitation. It also opens the door to applications that would take full advantage of the intrinsic properties of diamond in such areas as semiconductors, lasers, optics, opto-electronics, wear and corrosion coatings, and others.

To estimate a market would be no more than a guessing game since few applications at this time (1993) have reached the commercial stage and there is no historical background in patterns of growth and market size to buttress any predictions.

Some actual and potential applications of CVD diamond are listed in Table 13.7.[30]-[32]

In the following sections, several typical applications are described.

Table 13.7. Actual and Potential Applications of CVD Diamond

Grinding, cutting:	Inserts Twist drills Whetstones Industrial knives Circuit-board drills	Oil drilling tools Slitter blades Surgical scalpels Saws
Wear parts:	Bearings Jet-nozzle coatings Slurry valves Extrusion dies Abrasive pump seals Computer disk coatings	Engine parts Medical implants Ball bearings Drawing dies Textile machinery
Acoustical:	Speaker diaphragms	
Diffusion, corrosion:	Crucibles Ion barriers (sodium)	Fiber coatings Reaction vessels
Optical coatings:	Laser protection Fiber optics X-ray windows	Antireflection UV to IR windows Radomes
Photonic devices:	Radiation detectors	Switches
Thermal management:	Heat-sink diodes Heat-sink PC boards	Thermal printers Target heat-sinks
Semiconductor:	High-power transistors High-power microwave Photovoltaic elements	Field-effect transistors UV sensors

5.2   Grinding, Cutting, and Wear Applications

CVD-diamond coatings provide excellent protection against chemical and physical abrasion and wear due to their great hardness and wear resistance, good lubricating properties, and their general chemical inertness. As such, they may find a niche in cutting and grinding applications, where they would be essentially in direct competition with single-crystal diamonds (see Ch. 12, Sec. 5.2).

Single crystals have to be bonded to a substrate in a separate operation and that, coupled with their high cost, limits their general use in tooling (see Ch. 12). CVD-diamond coatings may provide a cheaper solution once the adhesion problem is solved satisfactorily. As mentioned in Sec. 4.5, polycrystalline materials do not show the wear anisotropy characteristic of single-crystal diamond and from that standpoint are better performers.

Substrate Materials. Cemented tungsten carbide (WC) is a major substrate cutting tool material. It has been successfully coated with CVD diamond with good adhesion especially with an intermediate carbide-former layer.[44]-[46]

Another tool material is silicon nitride which, when coated with CVD diamond, performed well in the machining of Al-Si alloys. These alloys are very abrasive and are usually machined with bonded polycrystalline diamond (PCD) (see Ch. 12, Sec. 5.2). The diamond-coated inserts have the advantage due to the availability of multiple edges on each insert.[46]

The high deposition temperature ($\sim 1000°C$) of CVD diamond may exclude deposition on substrates such as high-speed steels for drill bits and cutting tools.[48] These steels would lose their hardness and strength during processing and in order to coat them, a process must be developed that will keep the deposition temperature below 600°C.

5.3   Thermionic Applications

A diamond-coated cylinder of refractory metal (Nb, Mo, W, or Re) may replace the cylindrical ceramic insulators used in thermionic fuel elements (TFE) for space nuclear-power systems. A TFE converts the thermal energy from a nuclear reaction into electrical energy and is designed to provide power to a spacecraft for as long as ten years. The unique properties of diamond would prevent power loss due to electrical leakage while simultaneously removing the waste heat from the fuel element. A molybdenum cylinder, 16 mm in diameter and 75 mm in length, was coated with 5 μm of CVD diamond by DC plasma.[49][50] Its performance under TFE operating conditions is being evaluated.

5.4 Electronic and Semiconductor Applications

With its excellent semiconductor properties, good radiation and temperature resistance, and high thermal conductivity, diamond is the ideal material for many semiconductor applications such as high-power and high-frequency transistors and cold cathodes, or in the harsh environment found in internal-combustion and jet engines. With the advent of CVD diamond, it is now possible to take advantage of these properties.

Diamond, however, is not the universal semiconductor panacea: it is an *indirect bandgap* semiconductor and does not lase. In addition, present semiconductor materials such as silicon and gallium arsenide are solidly entrenched with a well-established technology and competing with them will not be an easy task. CVD diamond will also compete with silicon carbide which has also an excellent potential as a high-performance semiconductor material and is considerably easier and cheaper to produce.[7]

Many problems must be solved before practical, reliable, and cost-effective diamond semiconductor devices become available. Yet, the prospects are good, particularly if epitaxial single-crystal or highly oriented polycrystalline diamond can be effectively deposited.[51]

Preliminary developments of semiconductor devices based on CVD diamond are already under way including field-effect transistors (FET) which are proving superior to silicon devices and are characterized by high-power handling capacity, low saturation resistance, and excellent high-frequency performance.[28]

5.5 Thermal Management (Heat Sink) Applications

The design of integrated circuits (ICs) has reached the point where one million or more components can be put on a single chip (known as very large scale integration or VLSI) and higher densities will soon be needed as circuit designers are constantly demanding higher performance, expanded memory, faster access time, and shrinking geometry which now reaches the submicron size.

The density of these integrated circuits (ICs), especially those found in microwave, millimeter-wave, and opto-electronic devices such as laser diodes and laser-repeater assemblies, is presently limited by the large amount of heat generated by the extremely close packing of the electronic components on the chip and the high frequencies and power levels. To remove this heat, it is often necessary to use hybrid circuits and bulky heat-dissipation devices or complicated and expensive refrigeration since present heat-sink materials are no longer adequate.

Diamond, an electrical insulator with the highest thermal conductivity at room temperature, provides a solution. It compares favorably with conventional heat-sink materials such as aluminum, copper, or ceramics such as beryllia and aluminum nitride as shown in Table 13.8.[39][52][53] Like these two ceramic materials, diamond is an excellent electrical insulator and no electrical insulation between the IC board and the heat-sink is required. A diamond heat-sink should allow clock speeds greater than 100 GHz compared to the current speeds of less than 40 GHz.

Table 13.8. Properties of Heat-Sink Materials

Properties	AlN	BeO	CVD Diamond
Resistivity, ohm·cm	10^{14}	10^{14}	10^{12} - 10^{16}
Dielectric constant @ 1 MHz	8.8	6.7	5.6
Dielectric strength, kV/mm	10^5	10^5	10^6
Thermal conductivity, W/m·K	140 - 220	260	1700
Specific heat, J/g·K	0.7	1.0	0.51
Thermal expansion (0 - 400C), $10^{-6}/°C$	4.1	7.2	2.0

It should be noted that beryllia is being phased out because of the acute health problems associated with its use.

Heat sinks in the form of thin slices prepared from single-crystal natural diamond are already used commercially but are limited in size to approximately 3 x 3 x 1 mm. These single-crystal diamonds are gradually being replaced by CVD diamond which is now available in shapes up to 15 cm in diameter.[30][31] CVD-diamond heat sinks may be at the threshold of large-scale commercialization and could be the most promising application of the material.[39]

5.6   Optical Applications

A CVD diamond coating is yet to be produced with the optical clarity of single-crystal diamond and, for that reason, CVD diamond has found little application so far in optics.   Typically CVD diamond coatings are translucent. Some are nearly white, some have a black tinge.[54]

The rough surface of CVD diamond is also a drawback since an optically smooth coating is an essential requirement for most optical applications. The surface can be smoothed to some degree by increasing the nucleation density and by deposition strongly <100> textured coatings where the surface consists of flat {100} faces more or less parallel to the substrate.[55]

Diamond for IR Windows. A new optical-brazing process bonds a thin layer of CVD diamond on a zinc sulfide or zinc selenide IR window. This thin layer is obtained by depositing diamond on a silicon substrate and dissolving the silicon. The diamond layer is then brazed to the IR window with the smooth interface surface on the outside. The brazing material is an index-matching chalcogenide glass.[56]   This process is a possible solution to the optical-roughness problem mentioned above. The process is shown schematically in Fig. 13.9. In such an application, diamond would compete directly with diamond-like carbon (DLC) (see Ch. 14, Sec. 5.2).

Advances in deposition techniques have now made possible the fabrication of free-standing shapes such as the 7.6 cm (3") diameter hemispherical dome for IR-window applications shown in Fig. 13.10 which may eventually replace the present IR materials.

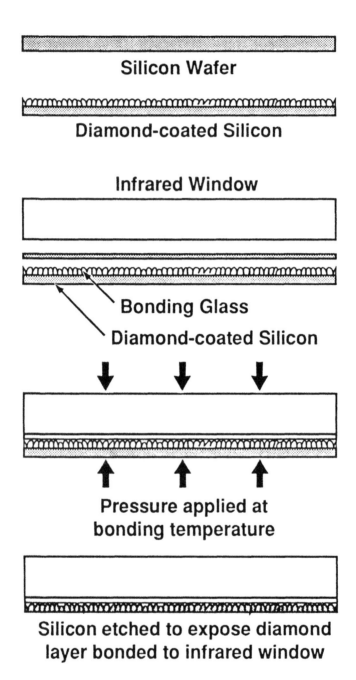

Silicon Wafer

Diamond-coated Silicon

Infrared Window

Bonding Glass

Diamond-coated Silicon

Pressure applied at
bonding temperature

Silicon etched to expose diamond
layer bonded to infrared window

Figure 13.9. Schematic of optical brazing process for bonding a CVD-diamond layer to an IR window.[56]

Figure 13.10. Free-standing CVD diamond dome, 7.6 cm diameter, for IR window applications. *(Photograph courtesy of Norton Diamond Film, Northboro, MA.)*

Diamond for X-ray Windows. CVD diamond is an excellent material for windows of x-ray spectrometers. In thin cross section, it is transparent to the x-rays generated by low-energy elements such as boron, carbon, and oxygen (see Ch. 11, Sec. 7.0, and Fig. 11.14) and is superior to beryllium which is the present standard material. Such a window is now commercially available and is shown schematically in Fig. 13.11.[57] It is leak-tight to helium and capable of withstanding at least one atmosphere differential pressure.

X-Ray Lithography. CVD-diamond coatings are considered for x-ray lithography masks[58] (see Ch. 14, Sec. 5.3).

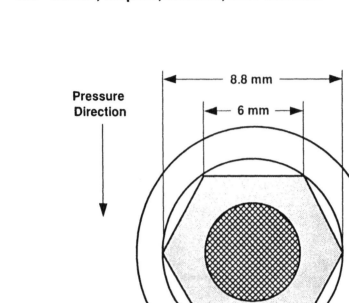

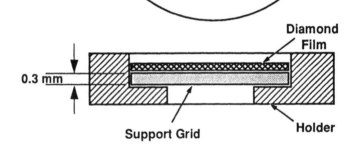

Figure 13.11. Schematic of CVD-diamond x-ray window assembly.[57]

REFERENCES

1. Eversole, W. G., "Synthesis of Diamond," U.S. Patent No. 3,030,188 (Apr. 17, 1962)

2. Angus, J. C., Will, H. A., and Stanko, W. S., *J. Appl. Phys.*, 39:2915-2922 (1968)

3. Derjaguin, B. V. and Fedoseev, D. V., *Scientific American*, 233(5):102-109 (1975)

4. Derjaguin, B. V. and Fedoseev, D. V., *Jzd. Nauka* (in Russian), Ch. 4, Moscow (1977)

5. Spitzyn, B. V., Bouilov, L. L., and Derjaguin, B. V., *J. of Crystal Growth*, 52:219-226 (1981)

6. Spear, K., *J. Amer. Ceram. Soc.* 7(2):171-91 (1989)

7. Pierson, H. O., *Handbook of Chemical Vapor Deposition*, Noyes Publication, Park Ridge, NJ (1992)

8. Yarborough, W. A., *J. Mater. Res.*, 7(2):379-383 (1992)

9. Yarborough, W. A., *Diamond Films and Technology*, 1(3):165-180 (1992)

10. Yasuda, T., Ihara, M., Miyamoto, K., Genchi, Y., and Komiyama, H., *Proc. 11th. Int. Conf. on CVD*, (K. Spear and G. Cullen, eds.) 134-140, Electrochem. Soc., Pennington, NJ (1990)

11. Frenklash, M. and Spear, K., *J. of Mat. Res.*, 3(1):133-1406 (1988)

12. Garrison, B. J., Dawnhashi, E., Srivastava, D., and Brenner, D. W., *Science*, 255:835-838 (1992)

13. Yarborough, W. A., *Applications of Diamond Films and Related Materials*, (Y. Tzeng, et al., eds.), 25-34, Elsevier Science Publishers (1991)

14. Setaka, N., *Proc. 10th. Int. Conf. on CVD*, (G. Cullen, ed.), 1156-1163, Electrochemical Soc., Pennington, NJ (1987)

15. Hsu, W. L., *J. Vac. Sci. Technol.*, A(6)3:1803-1811 (May/June 1988)

16. Saito, Y., Sato, K., Tanaka, H., Fujita, K., and Matsuda, S., *J. Mat. Sci.*, 223(3):842-46 (1988)

17. Patterson, D. E., et al., *Applications of Diamond Films and Related Materials*, (Y. Tzeng, et al., eds.), 569-576, Elsevier Science Publishers (1991)

18. Hong, F. C., Liang, G., Chang, D., and Yu, S., *Applications of Diamond Films and Related Materials*, (Y. Tzeng, et al., eds.), 577-828, Elsevier Science Publishers (1991)

19. Rudder, R. A., et al., *Applications of Diamond Films and Related Materials*, (Y. Tzeng, et al., eds.), 583-588, Elsevier Science Publishers (1991)

20. Kadono, M., Inoue, T, Miyanaga, A., and Yamazaki, S., *App. Phys. Lett.*, 61(7):772-773 (17 Aug. 1992)

21. Thorpe, M., *Chemical Engineering Progress*, 43-53 (July 1989)

22. Bachmann, P. K., Gartner, G., and Lydtin, H., *MRS Bulletin*, 52-59 (Dec. 1988)

23. Kearney, K., *Semiconductor International,* 66-68 (March 1989)

24. Ohtake, N. and Yashikawa, M., *J. Electrochem. Soc.,* 137(2):717-722 (1990)

25. Kurihara, K., Sasaki, K, Kawarada, M., and Koshino, M., *App. Phys. Lett.*, 52(6):437-438 (1988)

26. Yamamoto, M., et al., U.S. Patent No. 4,851,254 (Jul. 25, 1989)

27. Woodin, R. L., Bigelow, L. K., and Cann, G. L., *Applications of Diamond Films and Related Materials*, (Y. Tzeng, et al., eds.), 439-444, Elsevier Science Publishers (1991)

28. Fujimori, N., *New Diamond*, 2(2):10-15 (1988)

29. Matsumoto, S., *Proc. of the Mat. Res. Soc.,* 119-122, Spring Meeting, Reno, NV, Extended Abstract (April 1988)

30. *Norton CVD Diamond,* Technical Bulletin from Norton Co., Northboro, MA (1992)

31. *Diamond Coating, A World of Opportunity,* Technical Brochure, Genasystems Inc., Worthington, OH (1991)

32. Pickrell, D. J. and Hoover, D. S., *Inside ISHM,* 11-15 (July/Aug. 1991)

33. Schafer, L., Sattler, M., and Klages, C. P., *Applications of Diamond Films and Related Materials*, (Y. Tzeng, et al., eds.), 453-460, Elsevier Science Publishers (1991)

34. Sommer, M. and Smith, F. W., *J. of Materials Research*, 5(11):2433-2440 (1990)

35. Yarborough, W. A., Stewart, M. A., and Cooper, J. A., Jr., *Surface and Coatings Technology*, 39/40:241-252 (1989)

36. Hirose, Y., *Applications of Diamond Films and Related Materials*, (Y. Tzeng, et al., eds.), 471-472, Elsevier Science Publishers (1991)

37. Anthony, T. R. and Banholzer, W. F., *Diamond and Related Materials*, I:717-726 (1992)

38. Anthony, T. R., et al., *Physical Review B*, 42(2):1104-1111 (July 1990)

39. Seal, M., *Applications of Diamond Films and Related Materials*, (Y. Tzeng, et al., eds.), 3-7, Elsevier Science Publishers (1991)

40. Joffreau, P. O., Haubner, R., and Lux, B., *Proc. MRS Spring Meeting*, Reno, NV (Apr. 1988)

41. Rudder, R. A., et al., *Diamond Optics*, 969:72-78, SPIE (1988)

42. Sioshansi, P., "Polishing Process for Refractory Materials," U.S. Patent No. 5,154,023 (assigned to Spire Corp.) (Oct. 13, 1992)

43. Graebner, J. E. and Herb, J. A., *Diamond Films and Technology*, 1(3):155-164, MYU, Tokyo (1992)

44. Shibuki, K., Yagi, M., Saijo, K., and Takatsu, S., *Surface and Coating Technology*, 36:295-302 (1988)

45. Murakawa, M., Takuchi, S., Miyazawa, M., and Hirose, Y., *Surface and Coatings Technology*, 36:303-310 (1988)

46. Kikuchi, N., Eto, H., Okamura, T., and Yoshimura, H., *Applications of Diamond Films and Related Materials*, (Y. Tzeng, et al., eds.), 61-68, Elsevier Science Publishers (1991)

47. Soderberg, S., Westergren, K., Reineck, I., Ekholm, P. E., and Shashani, H., *Applications of Diamond Films and Related Materials*, (Y. Tzeng, et al., eds.), 43-51, Elsevier Science Publishers (1991)

48.   Chen, H., Nielsen, M. L., Gold, C. J., and Dillon, R. O., *Applications of Diamond Films and Related Materials*, (Y. Tzeng, et al., eds.), 137-142, Elsevier Science Publishers (1991)

49.   Adams, S. F. and Landstrass, M. I., *Applications of Diamond Films and Related Materials*, (Y. Tzeng, et al., eds.), 371-376, Elsevier Science Publishers (1991)

50.   Wu, R. L., Garacadden, A., and Kee, P., *Applications of Diamond Films and Related Materials*, (Y. Tzeng, et al., eds.), 365-370, Elsevier Science Publishers (1991)

51.   Geis, M. W., *Diamond and Related Materials*, I:684-687 (1992)

52.   Eden, R. C., *Applications of Diamond Films and Related Materials*, (Y. Tzeng, et al., eds.), 259-266, Elsevier Science Publishers (1991)

53.   Hoover, D. S., Lynn, S.- Y., and Garg. D., *Solid State Technology*, 89-91 (Feb. 1991)

54.   Kawarada, H., *Jpn. J. of App. Physics*, 27(4):683-6 (April 1988)

55.   Wild, C., Muller-Sebert, W., Eckermann, T., and Koidl, P., *Applications of Diamond Films and Related Materials*, (Y. Tzeng, et al., eds.), 197-205, Elsevier Science Publishers (1991)

56.   Partlow, W. D., Witkowski, R. E., and McHugh, J. P., *Applications of Diamond Films and Related Materials*, (Y. Tzeng, et al., eds.), 161-168, Elsevier Science Publishers (1991)

57.   Diamond X-Ray Window Product Data, Crystallume, Menlo Park, CA (1988)

58.   Hamilton, R. F., Garg. D., Wood, K. A., and Hoover, D. S., *Proc. X-Ray Conf.*, Vol. 34, Denver (1990)

14

Diamond-Like Carbon (DLC)

1.0 GENERAL CHARACTERISTICS OF DLC

As mentioned in Chs. 12 and 13, a new form of carbon coating is now available which is neither diamond nor graphite and is known as diamond-like carbon (DLC). The term DLC was introduced in 1973 by Aisenberg, a pioneer in this field.[1]

DLC can be considered as a metastable carbon produced as a thin coating with a broad range of structures (primarily amorphous with variable sp^2/sp^3 bonding ratio) and compositions (variable hydrogen concentration).[2][3] DLC coatings alter the surface properties of a substrate in a manner similar to that of CVD diamond (see Table 13.2, Ch. 13).

As opposed to diamond which occurs naturally and can be synthesized at both high and low pressure, DLC is not a natural material and can only be produced at low pressure. DLC has two disadvantages: low deposition rate and high internal stress. Unlike diamond, it cannot be obtained as thick monolithic shapes, at least with the present technology.

However, DLC has properties similar to CVD diamond and it is easier to process without the high-temperature substrate requirements and with little restriction on size. A number of important applications have been developed with a promising future.

2.0   STRUCTURE AND COMPOSITION OF DLC

2.1   Graphite, Diamond, and DLC

The differences between graphite, diamond, and DLC are summarized in Table 14.1.[6]

Table 14.1. Comparison of Graphite, Diamond, and DLC

	Graphite	Diamond	DLC
Composition	Pure carbon	Essentially carbon (<1 at.% hydrogen)	Up to 50 at.% hydrogen
Microstructure	Crystalline	Crystalline	Amorphous
Atom-bonding state	sp^2 only	sp^3 only	sp^2, sp^3, sp^1 (variable ratio)
Stability	Stable	Stable	Metastable
Raman spectrum	Sharp peak at 1580cm	Sharp peak at 1332 cm	Broad humps at 1330 & 1550 cm
Electrical conductivity	Conductor (*ab* direction)	Insulator	Insulator

2.2   Analytical Techniques

Several analytical techniques are used to characterized carbon materials such as diamond and DLC. These include x-ray diffraction, transmission electron microscopy, electron-energy-loss spectroscopy (EELS), and laser-Raman spectroscopy, the last two techniques being probably the most frequently used.

As mentioned in Ch. 11, Sec. 2.1, laser-Raman spectroscopy is used extensively to detect the diamond structure. It is also useful to characterize the near-surface of black, opaque materials such as graphite and DLC since the laser light is able to penetrate the surface of these materials a few tens of namometers. The Raman spectra are sensitive to disruption of translational symmetry such as occurs in small crystals and, as such, is particularly useful in the analysis of disorder and crystallite formation in DLC materials.[7]

2.3 Structure and Categories of DLC

A truly amorphous solid such as glass lacks any degree of crystallinity and has no lattice long-range order. Unlike glass, DLC is composed of small crystallites which have a local atomic configuration that can be either tetrahedral (sp^3) or planar threefold (sp^2) but are small enough that electron diffraction patterns indicate an amorphous material. DLC has no long-range order but forms a random stable network, similar to that of vitreous carbon (see Ch. 6, Sec. 3.1). As a result, DLC is generally considered to be amorphous. It has also been clearly demonstrated that the physical and chemical properties of the DLC films are closely related to the chemical bonding and the miscrostructure.[4][7][8]

DLC can be divided into two closely related categories known as amorphous DLC and hydrogenated DLC.

2.4 Amorphous DLC (a-C)

A sizeable proportion of the sites in amorphous DLC have the sp^3 coordination (diamond) in the form of individual or polycrystalline diamond in small clusters (5 - 10 nm), interspersed with amorphous regions. The sp^2 carbon sites are generally lacking as evidenced by the absence of π plasmon absorption shown by EELS.[4] The hydrogen content is less than one atomic percent and a small amount of argon may remain trapped in the lattice.[2] The exact overall structure is still uncertain. The material is generally produced by sputtering from a solid carbon target (see Sec. 3.2) and is usually referred to as "a-C".

2.5 Hydrogenated DLC (a-C:H or H-DLC)

Hydrogenated DLC is also considered amorphous but, unlike a-C, it contains a variable and appreciable amount of hydrogen (up to fifty atomic percent). Its structure consists of an essentially amorphous network with isolated clusters dominated by the sp^2 configuration (graphite) with some sp^3 (diamond). Hydrogen is believed to play an essential role in determining the bonding configuration by helping to form the sp^3 bond, probably in a manner similar to the formation of CVD diamond (see Ch. 13). The exact location of the hydrogen atoms is still conjectural but hydrogen, being monovalent, can serve only as a terminating atom of the carbon network.[5]

The ratio of sp^3 to sp^2 varies considerably as a function of the hydrogen content as shown in Fig. 14.1.[5] There is also some evidence of trigonal, triple carbon bonding (sp^1) (see Ch. 2, Secs. 3.0 and 4.0).

The material is produced by plasma action in a hydrocarbon atmosphere (see Sec. 3.3) and is referred to as hydrogenated DLC (H-DLC or a-C:H).

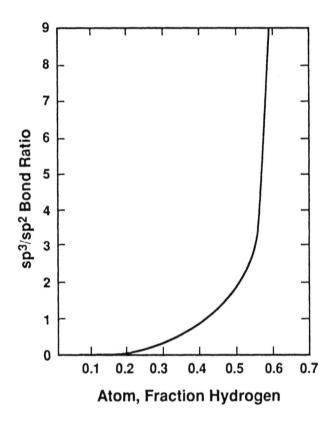

Atom, Fraction Hydrogen

Figure 14.1. Ratio of sp^3 to sp^2 hybrid bonds in diamond-like carbon (DLC) as a function of hydrogen content.[5]

3.0 PROCESSING OF DLC

3.1 Processing Characteristics

Unlike the activation of the CVD-diamond deposition which is a chemical process, the activation of DLC deposition is a physical phenomenon in which the sp^3 configuration is stabilized by ion bombardment of the growing film instead of (or in addition to) atomic hydrogen. The process is known in general terms as physical vapor deposition (PVD).

Deposition Model. The formation of DLC is still conjectural but a proposed mechanism can be summarized as follows.[9] The conversion of graphite to diamond requires a compression from the lower energy and density of graphite to the higher energy and density of diamond as the coordination shift from sp^2 to sp^3. This may happen under the impact of an energetic particle, such as argon or carbon itself, generated by an ion beam, laser or other means. The optimum C^+ kinetic energy range is 30 - 175 eV.[10] Higher energy may drive the atom back to the sp^2 configuration unless the particle is rapidly quenched. This quenching is normally achieved with a high thermal diffusivity substrate that is kept cool.

In the case of hydrogenated DLC (a-C:H), it is believed that both ion bombardment and the presence of atomic hydrogen are the key factors in promoting the sp^3 configuration.[2]

Process Classification. The DLC deposition processes are divided into two broad categories:

1. The purely PVD processes which include ion beam sputtering, laser or glow discharge from a solid carbon target (producing a-C).

2. The PVD/CVD processes such as DC or RF plasma or ion-beam deposition from a gaseous hydrocarbon (producing a-C:H).

3.2 DLC by PVD Processes from a Carbon Target

Ion-Beam Sputtering. The activation process for the deposition of a-C DLC is based on the well-established industrial process of ion-beam sputtering.[11][12] A solid carbon cathode, also known as the target, is the carbon source. The deposition mechanism is summarized as follows:

1. A gas, usually argon, exposed to a high voltage, becomes ionized and produces a low-pressure electric discharge (the glow discharge described in Ch. 13, Sec. 3.2).

2. The target is negatively biased and thus attracts the argon ions which are accelerated in the glow discharge. The resulting bombardment by the energetic ions causes the sputtering of carbon atoms which are ejected by momentum transfer between the incident ion and the target, as shown schematically in Fig. 14.2.[2]

3. The sputtered carbon atoms are energetic and move across the vacuum chamber to be deposited on the substrate. The latter is placed so that it can intercept these atoms with maximum efficiency as shown in the deposition apparatus of Fig. 14.3. In this system, a second ion source bombards the substrate to enhance the deposition mechanism.

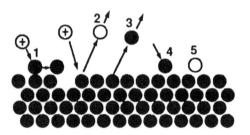

1. Impacting gaseous ion moving surface atom in advantageous site

2. Gaseous impurity atom sputtered away from surface

3. Loosely bonded solid atom sputtered away from surface

4. Solid atom or ion deposited on surface

5. Gaseous impurity atom remaining on surface

Figure 14.2. Effect of impacting energetic ions on surface carbon atoms.[2]

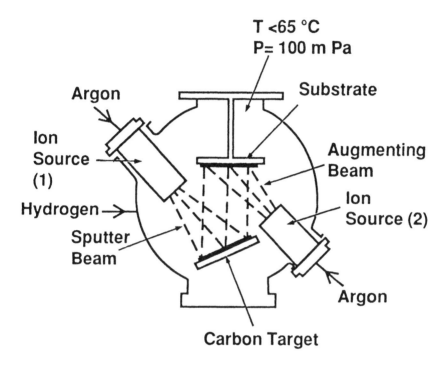

T <65 °C
P= 100 m Pa

Figure 14.3. Schematic of dual ion-beam apparatus for the deposition of diamond-like carbon (DLC).

Ionized Species. Argon is the preferred bombarding gas because it has a higher atomic weight and is easier to ionize than the other two common inert gases, neon and helium. A higher atomic weight gives higher sputtering yield especially if the weight of the bombarding particle is greater than that of the target atom (the atomic mass of argon is thirty-nine times that of carbon-12). As sputtering proceeds, the carbon target becomes gradually depleted and eventually will disappear altogether.

Sputtering Yield and Deposition Rate. The sputtering yield of carbon, i.e., the number of atoms ejected from the target surface per incident ion, is very low (0.12 when bombarded with argon ions). Since the sputtering yield is the major factor in determining the deposition rate of the sputtered coating, it follows that the sputtered deposition of carbon is inherently a slow process, suitable for thin coatings (<1 μm) but not practical for thicker coatings, at least at this stage of technology.

Ion-Beam Sputtering Process and Equipment. Ion-beam sputtering requires low pressure to remove all traces of background and contaminant gases which could degrade the coating. This is achieved by cryogenic pumps capable of producing a vacuum of 10^{-5} Pa with good pumping speed. After evacuation, the system is refilled with argon to a partial pressure of 0.1 to 10 Pa. Higher pressure would not allow the ions and ejected atoms to travel relatively unimpeded by collision with other atoms or molecules. In other words the mean-free-path would be too short.

Magnetron Sputtering. The development of a magnetically enhanced cathode or *magnetron* has considerably expanded the potential of sputtering. The magnetron sends the electrons into spiral paths which increases collision frequency and enhances ionization.[7][11] A typical DC magnetron system for the deposition of DLC has the following characteristics:[13]

Discharge power: 900 W

Argon working pressure: 0.4 Pa

Substrate potential: 0 - 100 V

DLC deposition rate: 0.25 μm/h

Carbon particle: mostly neutral atoms

Argon atoms: mostly highly energetic ions

Variations of the basic sputtering system include the following.

1. A 100 mm graphite-disk target is bombarded with an ion beam of 25 mm diameter with 1200 eV ion energy and 60 mA ion current and ion-current density of 1 mA/cm² at the target and 0.04 mA/cm² at the substrate.[15] The DLC deposition rate is 0.3 to 0.4 μm/h. Diamond structure is not observed in the film unless hydrogen ions are added to the argon ions. With hydrogen present, some diamond structures are evident. However many non-diamond configurations remain, and the coating is brownish and electrically conductive. These hydrogen ions appear to play the same role as atomic hydrogen in CVD deposition.

2. Two opposite (facing) targets and a confined argon plasma between the targets are placed in a magnetic field of 160 Gauss as shown in Fig. 14.4.[16] Applied voltage is 750 V and discharge current is 0.4 amp. The bias voltage to the substrate is applied with a 13.56 MHz power source, and a-C films in the range of 0.05 to 0.2 μm are deposited at argon pressure in the range of 0.05 to 1.3 Pa.

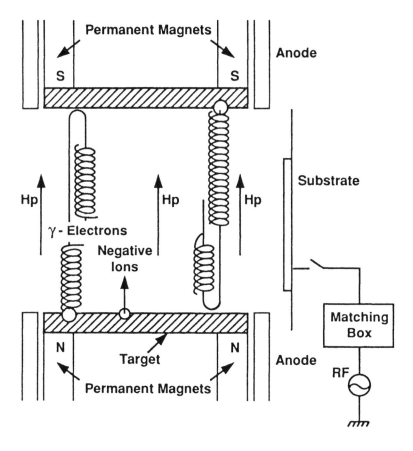

Figure 14.4. Schematic of facing-targets sputtering apparatus for the deposition of diamond-like carbon (DLC).[16]

3.3   DLC by PVD-CVD Process from a Hydrocarbon Source

In the PVD-CVD process, the carbon source is a hydrocarbon gas instead of the solid carbon target of the purely PVD process. The same activation methods (ion-beam sputtering, laser, glow-discharge, or others) are used but a-C:H is deposited instead of a-C.

Deposition by RF Activation. A common activation method is a high-frequency RF gas discharge (13.56 MHz), generated in a mixture of hydrogen and a hydrocarbon such as methane (CH_4), n-butane (C_4H_{10}), or acetylene (C_2H_2). A diagram of the equipment is shown in Fig. 14.5.[17] Two factors, the asymmetry of the electrodes and the considerable difference in mobilities between electrons and ions, cause the spontaneous generation of a negative potential at the substrate which, as a result, is bombarded by the ionized gas species.

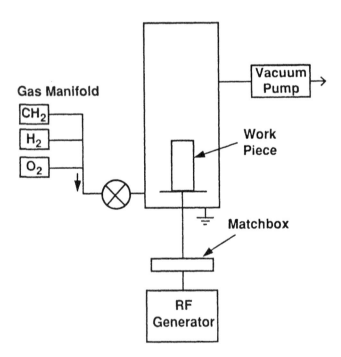

Figure 14.5. Schematic of apparatus for the deposition of a-C:H DLC.[17]

The resulting coating can be a-C:H or a soft polymer-like graphite depending on the applied energy. Deposition rate is 0.5 to 2 μm/hr. A clean deposition surface is obtained by chemically etching the substrate followed by sputter cleaning with argon just prior to the actual deposition.

Since a solid carbon target is not needed and the carbon source is a gas, greater flexibility is possible in the positioning and geometry of the substrate(s) as a coating is deposited on every surface exposed to the gas (as opposed to sputtering which is essentially a line-of-sight process). Unlike the purely PVD process, large parts can be coated (as long as they can be electrically contacted) with present production equipment.[17]

Deposition by Ion-Beam Activation. A typical ion-beam activated system has a 30 cm hollow-cathode ion source with its optics masked to 10 cm. Argon is introduced to establish the discharge followed by methane in a 28/100 ratio of methane molecules to argon atoms. The energy level is 100 eV, the acceleration voltage is 600 V, and the resulting deposition rate is 0.5 μm/h.

A similar system has a dual ion-beam. A primary beam sputters carbon while the growing film is being simultaneously bombarded with argon ions generated from a second ion source.[18] Another system is based on a microwave discharge generated by electron cyclotron resonance (ECR).[19][20] The principle of ECR is described in Ch. 13, Sec. 3.3.

3.4 Substrate Heating

Under normal circumstances during deposition by high-frequency discharge, the substrate remains at low temperature (<300°C) and a wide variety of materials can be coated including plastics. This characteristic is a major advantage of DLC processing over CVD diamond.

4.0 CHARACTERISTICS AND PROPERTIES OF DLC

4.1 Summary of Properties

Graphite and diamond are materials with a well-defined structure and properties which vary within a relatively narrow range of values. DLC is different as its structure and composition may vary considerably and, as a result, so do some of its properties. This is not necessarily a disadvantage since it is often possible to control and tailor these properties to fit specific applications (for instance, the index of refraction). The properties of DLC are generally similar to those of CVD diamond but different in some key areas as reviewed below. Like CVD diamond, DLC is a recently developed material, only available as a thin coating. This makes property measurement a difficult task due the uncertain effect of the substrate. This must be taken into account in evaluating the values reported in the literature.[7][13][21][22]

The properties of DLC are summarized and compared with those of CVD diamond in Table 14.2 (see also Table 13.5 in Ch. 13).

Table 14.2. Properties of DLC and CVD Diamond Coatings

	CVD Diamond	DLC
Density, g/cm^3	3.40 ± 0.10	1.8 - 2.8
Thickness range, μm	1 - 1000	0.1 - 5
Internal stress	Tensile (moderate)	Compressive (high) 1.3 - 1.6 GPa
Thermal conductivity at 25°C, W/m·K	>1300	400 - 1000
Bangap, eV	5.48	0.8 - 3
Index of refraction @ 10 μm	2.34 - 2.42	1.8 - 2.4
Electrical resistivity, Ω·cm	10^{12} - 10^{16}	10^5 - 10^{15}
Dielectric constant (45 MHz - 20 GHz)	5.6	8 - 12
Vickers hardness, kg/mm^2	5000 - 10000	2000 - 9000
Coefficient of friction*	0.05 - 0.15	0.01 - 0.3

* Varies with humidity

4.2 Internal Stress and Adhesion

At the present, it is not possible to deposit thick DLC coatings as they tend to delaminate and separate from the substrate when the thickness is greater than a few microns. This is the result of high internal compressive stresses (1.3 - 1.6 GPa) which appear to be related to the hydrogen content of the material.[3][4] Thus the a-C:H coatings with their high hydrogen content are more highly stressed than a-C coatings which have little hydrogen.

Generally, the adhesion of DLC to its substrate is satisfactory, providing that suitable cleaning by ion bombardment is achieved prior to deposition. Adhesion to carbide formers such as Si, Ge, Mo, Ti, W, and iron alloys is particularly good. Adhesion on silicide formers is improved by depositing an intermediate layer of Si.[3] Adhesion to various substrates is shown in Table 14.3.[2] The test method is the Sebastian adhesion tester. Coating thickness averages 10 nm. The adhesion of a-C and a-C:H is generally similar.

Table 14.3. Adhesion of DLC to Various Substrates

Substrate	Adhesion (MPa)
Silicon {111} and {100}	55
Polished stainless steel	34
Most metals	~ 30
Fused silica *	17
CR-39 optical plastic *	13

*Coating adhesion exceeded cohesive strength of substrate

4.3 Coating Morphology, Porosity, and Diffusional Property

An outstanding characteristic of DLC coating is its nanolevel smoothness and freedom of pinholes and other similar defects. From that standpoint, DLC is different from CVD diamond (see Ch. 13, Sec. 3.10). A

DLC coating planarizes the surface, i.e., the coating surface is smoother than the substrate, with surface roughness measured at 0.28 nm for a 100 nm coating thickness.[3] The material is an excellent diffusion barrier to sodium, moisture, and most gases.[2]

4.4   DLC/Graphite Transformation

DLC converts to graphite at lower temperature than diamond. An initial transformation has been observed as low as 250°C. The transformation is rapid at 400°C and proceeds by loss of hydrogen and subsequent graphitization. A maximum long-term use temperature for DLC is 250 - 300°C.

4.5   Optical Properties

DLC coatings are generally optically transparent in the wavelength range of 2.5 - 18 μm but may have a dark or yellow appearance. IR absorption in the 1 - 10 μm range is very low. The index of refraction of DLC varies from 1.8 to 2.4 at a 0.589 μm wavelength, decreasing with increasing amount of hydrogen (see Sec. 5.2).

4.6   Electrical Properties

As opposed to diamond, DLC has a variable electrical conductivity, which is a function of hydrogen content. It may not be a suitable semiconductor material since it has a relatively low bandgap, generally low resistivity, and low operating temperature, although semiconducting properties have been reported.[19]

4.7   Hardness

An outstanding property of DLC is its hardness. Vickers hardness ranges from 2000 to 9000 kg/mm^2. The large spread is due in part to the difficulty of testing thin coatings by indentation such as the Vickers test, since it is difficult to eliminate the substrate effect. Hardness also varies with the structure and composition.

5.0 APPLICATIONS OF DLC

DLC coatings have recently reached the production stage with applications in wear and erosion protection and in optics. The cost is generally similar to that of carbide or nitride films deposited by CVD or PVD techniques.

5.1 DLC in Wear and Tribological Applications

The high hardness, low coefficient of friction, and general chemical inertness of DLC coatings make them well-suited for applications involving wear and friction. In addition, these coatings have a very smooth surface and can be deposited with little restriction of geometry and size. These are important advantages and DLC will likely gain a foothold in the hard coating industry which is at the present dominated by the carbides and nitrides (TiC, TiN, Ti(CN) and WC).[6][17]

The major drawback of DLC is it lack of high-temperature resistance and thickness limitation which restricts its applications in tool coating. In this respect, DLC is similar to CVD diamond (see Ch. 13, Sec. 5.2).

In tribological applications, DLC coatings have good potential, particularly in non-lubricated conditions and in vacuum environment. These coatings, when rubbed against themselves, have a remarkably low coefficient of friction (0.02 - 0.04) and, because of their superior hardness, are much more wear resistant than Teflon™ or molybdenum disulfide. In addition, DLC coatings are very smooth in the as-deposited condition, which is another advantage in friction applications.

DLC coatings are already found in a number of applications, either on a development basis or in preliminary production. These applications include textile machinery, bearing surfaces, measuring instruments, air bearings, precision tooling, gears, fluid engineering systems, engine components, nozzles, and rotating equipment.[17] Some typical wear applications are as follows:

- DLC coatings for video tapes and high-density magnetic-recording disks are applied by sputtering with facing targets. The disks are made of polyvinyl acetate. The coatings are extremely thin and smooth and allow the smallest-possible clearance between the recording head and the surface of the disk.[3][7][23]

- DLC films on smooth metal substrates have a shiny appearance with a charcoal-black coloration and are hard, do not scratch, and are chemically resistant. They are being tested for decorative and protective applications such as jewelry and watch parts.[18]

5.2 Optical Applications of DLC

DLC is not as good an optical material as single crystal diamond but, as seen above, it has many advantages as a thin coating and is satisfactory in many applications. The following is a review of present and potential optical applications of DLC.

DLC in Infrared (IR) Optics. In Ch. 11, Sec. 6.2, it was shown that diamond is transparent in the infrared beyond 7 μm. DLC, although not as transparent as diamond, is suitable and is used or considered for a number of IR applications.[2][24]

DLC (a-C:H) provides an anti-reflection (AR) coating with an adjustable index of refraction which varies with the hydrogen content as mentioned above and can be produced to match any specific optical design. DLC coatings are particularly well-suited to germanium windows with 90% average transmission in the 8 - 12 μm wavelength range and to zinc-sulfide windows where they provide a reflectance value of 0.4% average in that same range.[25] Adhesion to the zinc sulfide is improved by an intermediate film of germanium.[26]

DLC coatings also have an opto-protective function. Infrared window materials such as germanium, magnesium fluoride, cadmium telluride, zinc sulfide, and zinc selenide are relatively soft and easily damaged and eroded by wind, rain, or particle impact. They have also poor resistance to corrosive environments. DLC coatings offer good protection with adequate optical properties. However, their narrow IR bandpass may limit the range of applications.

DLC Coatings for Laser Optics. Laser window materials must be capable of withstanding high levels of fluence (to 100 kW/cm^2 or more) and tight bypass-band specifications. Tests carried out on a silicon substrate coated with DLC (with measurable amount of sp^3 bonds) showed that the material has a high laser-damage threshold and is appreciably less damaged than common optical-coating materials. These findings established the suitability of DLC for high-power laser windows.[2][4]

5.3   DLC Coatings in Lithography

X-ray lithography is necessary to produce submicron patterns and gratings for integrated circuits (ICs). A major problem of this new technology is the development of a suitable mask. This has been solved, at least partially, by using DLC films (a-C;H) which have performed better than any other materials. These films form the bottom-layer etch mask below an electron-beam resist in a bilayer system. DLC is readily etched with oxygen reactive-ion etching and has a low-level etch rate with the dry etch used for the Si or GaAs substrates. The high mechanical integrity of DLC permits high-aspect ratios. With this technique, patterns as small as 40 nm have been transferred to the substrate.[3][27]

5.4   Miscellaneous DLC Applications

DLC coatings are found in the following applications:

- Biomedical: Coatings for hip joints, heart valves and other prostheses. DLC is biocompatible and blood compatible.[28]

- Biochemical: Coating for tissue culture flasks, microcarriers, cell culture containers, etc.

- Acoustical: DLC-coated tweeter diaphragm for dome speaker.[29]

5.5   Summary

DLC coatings are already in production in several areas (optical and IR windows) and appear particularly well-suited for abrasion and wear applications due to their high hardness and low coefficient of friction. They have an extremely smooth surface and can be deposited with little restriction of geometry and size (as opposed to CVD diamond). These are important advantages and DLC coatings will compete actively with existing hard coatings such as titanium carbide, titanium nitride, and other thin film materials. The major drawback of DLC is it lack of high temperature resistance which may preclude it from cutting- and grinding-tool applications, and limitations in thickness to a few microns due to intrinsic stresses.

REFERENCES

1. Aisenberg, S. and Chabot, R., *J. Appl. Phys.*, 42:2953 (1971)

2. Aisenberg, S. and Kimock, F. M., *Materials Science Forum,* 52&53:1-40, Transtech Publications, Switzerland (1989)

3. Grill, A., Patel, V., and Meyerson, B. S., *Applications of Diamond Films and Related Materials,* (Y. Tzeng, et al., eds.), 683-689, Elsevier Science Publishers (1991)

4. Angus, J. C., et al., *Diamond Optics,* SPIE, 969:2-13 (1988)

5. Angus, J. C. and Jansen, F., *J. Vac. Sci. Technol.,* A6(3):1778-1785 (May/June 1988)

6. Pierson, H. O., *Handbook of Chemical Vapor Deposition (CVD),* Noyes Publications, Park Ridge, NJ (1992)

7. Tsai, H. and Bogy, D. B., *J. Vac. Sci. Technol.,* A 5(6):3287-3312 (Nov/Dec 1987)

8. Cho, N. H., et al., *J. Materials. Res.,* 5(11) (Nov. 1990)

9. Cuomo, J. J., et al., *Applications of Diamond Films and Related Materials,* (Y. Tzeng, et al., eds.), 169-178, Elsevier Science Publishers (1991)

10. Kasi, S. R., Kang, H., and Rabalais, J. W., *J. Vac. Sci. Technol.,* A 6(3):1788-1792 (May/June 1988)

11. Wasa, K. and Hayakawa, S., *Handbook of Sputter Deposition Technology,* Noyes Publications, Park Ridge, NJ (1992)

12. Westwood, W. D., *MRS Bulletin,* 46-51 (Dec. 1988)

13. Richter, F., et al., *Applications of Diamond Films and Related Materials,* (Y. Tzeng, et al., eds.), 819-826, Elsevier Science Publishers (1991)

14. Deshpandey, C. V. and Bunshah, R. F., *J. Vac. Sci. & Tech.,* A7(3):2294-2302 (May-June 1989)

15. Kibatake, M. and Wasa, K., *J. Vac. Sci. Technol.,* A 6(3):1793-1797 (May/June 1988)

16. Hirata, T. and Naoe, M., *Proc. Mat. Res. Soc. Extended Abstract,* p. 49-51, Spring Meeting, Reno, NV (Apr. 1988)

17. Bonetti, R. S. and Tobler, M., *Amorphous Diamond-Like Coatings on an Industrial Scale,* Report of Bernex, Olten, Switzerland (1989)

18. Mirtich, M., Swec, D., and Angus, J., *Thin Solid Films*, 131:248-254 (1985)

19. Fujita, T. and Matsumoto, O., *J. Electrochem. Soc.*, 136(9):2624-2629 (1989)

20. Thorpe, M., *Chemical Engineering Progress*, 43-53 (July 1989)

21. *Diamond Coatings, a World of Opportunity,* Technical Brochure from Genasystems Inc., Worthington, OH (1991)

22. *Diamond-Like Carbon,* Technical Brochure from Ion Tech, Teddington, Middlesex, UK (1991)

23. Kurokawa, H., Nakaue, H., Mitari, T., and Yonesawa, T., *Applications of Diamond Films and Related Materials,* (Y. Tzeng, et al., eds.), 319-326, Elsevier Science Publishers (1991)

24. Lettington, A. H., *Applications of Diamond Films and Related Materials,* (Y. Tzeng, et al., eds.), 703-710, Elsevier Science Publishers (1991)

25. Cooper, *Photonic Spectra, 149-156 (Oct. 1988)*

26. Mirtich, M. J., *J. Vac. Sci. Technol.*, 44(6):2680-2682 (Nov/Dec 1986)

27. Horn, M. W., *Solid State Technology*, 57-62 (Nov. 1991)

28. Franks, J., "Member Implantable in Human Body," Basic EP Patent 302717 (890208), assigned to Ion Tech Ltd.

29. Yoshioka, T., Imai, O., Ohara, H., Doi, A., and Fujimori, N., *Surface and Coating Technology*, 36:311-318 (1988)

15

The Fullerene Molecules

1.0 GENERAL CONSIDERATIONS

1.1 State of the Art

The recent discovery of a family of large, solid carbon molecules with great stability, the so-called "fullerenes", has considerably extended the scope and variety of carbon molecules known to exist and is opening an entirely new chapter on the physics and chemistry of carbon, with many potential applications. The fullerenes can be considered as another major allotrope of carbon and its first stable, finite, discrete molecular form. They are different, in that respect, from the other two allotropes, graphite and diamond, which are not molecular but infinite-network solids. The other known carbon molecules, C_2 to C_{15}, are unstable and found only in the vapor phase at high temperature (see Ch. 2, Sec. 5.0).

The fullerenes are generally arranged in the form of a geodesic spheroid and thus were named after the inventor of the geodesic dome, the renowned architect Buckminster Fuller. They were originally (and still are occasionally) called "buckminster-fullerenes", a name fortunately shortened to fullerenes. They are also known as "buckyballs".

1.2 Historical Perspective

The fullerenes were discovered in 1985 by Smalley and Kroto who, while performing mass-spectroscopy analysis of carbon vapor, observed the presence of even-numbered clusters of carbon atoms in the molecular range of C_{30} - C_{100}, as illustrated in Fig. 15.1.[1]-[3] The experiment was carried out in the apparatus shown schematically in Fig. 15.2, consisting of a pulsed laser beam directed onto the surface of a rotating-translating graphite disk. Clusters of fullerenes were generated spontaneously in the condensing carbon vapor when the hot plasma was quenched by a stream of helium. This research provided the clue that led to the elucidation of the fullerene structure.

The practical synthesis of fullerenes as solid aggregates was demonstrated by Krätschmer and Huffman in 1990, by the simple evaporation of a graphite electrode as will be described in Sec. 5.0.[4] Fullerenes are now readily available in increasing quantities for study and evaluation.

These discoveries generated considerable interest in the scientific community and in the general public, as demonstrated by the number of publications on the subject, estimated to total nearly one thousand in 1993, only eight years after the fullerenes were first observed.

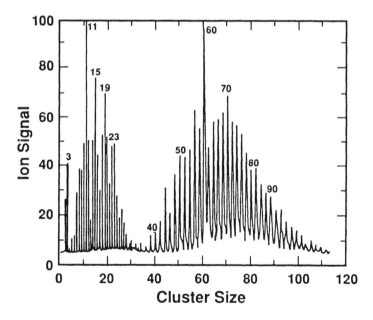

Figure 15.1. Time-of-flight mass spectrum of carbon clusters produced by laser vaporization of graphite.[2]

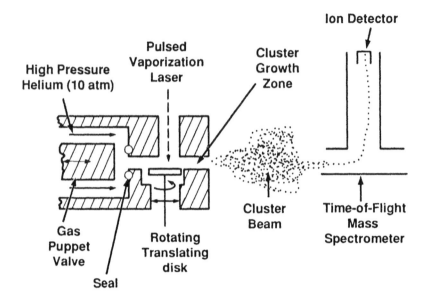

Figure 15.2. Carbon-cluster beam source with pulsed laser.[2][3]

2.0   STRUCTURE OF THE FULLERENE MOLECULES

2.1   Molecular Structure of the Fullerenes

Geodesic Structure. Unlike graphite or diamond, the fullerenes are not a single material, but a family of molecular, geodesic structures in the form of cage-like spheroids, consisting of a network of five-membered rings (pentagons) and six-membered rings (hexagons). In order to close into a spheroid, these geodesic structures must have exactly twelve pentagons, but can have a variable number of hexagons (m), with the general composition: C_{20+2m}. Such an even-numbered distribution is unique to the element carbon.

Atomic Bonding. In order to account for the bonding of the carbon atoms of a fullerene molecule, the hybridization must be a modification of the sp^3 hybridization of diamond and sp^2 hybridization of graphite (see Ch. 2, Secs. 3.0 and 4.0). It is such that the sigma (σ) orbitals no longer contain all of the s-orbital character and the pi (π) orbitals are no longer of the purely p-orbital character, as they are in graphite.

Unlike the sp^3 or sp^2 hybridizations, the fullerene hybridization is not fixed but has variable characteristics depending on the number of carbon atoms in the molecule. This number varies from twenty for the smallest geometrically (but not thermodynamically) feasible fullerene, the $C^{20,}$ to infinity for graphite (which could be considered as the extreme case of all the possible fullerene structures). It determines the size of the molecule as well as the angle Θ of the basic pyramid of the structure (the common angle to the three σ-bonds). The number of carbon atoms, the pyramidization angle (Θ - 90°), and the nature of the hybridization are related and this relationship (in this case the s character in the π-orbital) is given in Fig. 15.3.[5]

The bond lengths of the fullerenes are reported as 0.145 +/-0.0015 nm for the bonds between five- and six-membered rings, and 0.140 +/- 0.0015 nm for the bond between the six-membered rings.[6]

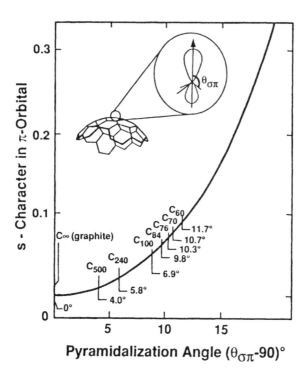

Figure 15.3. Hybridization of fullerene molecules as a function of pyramidization angle ($\Theta\sigma\pi$ - 90°). $\Theta\sigma\pi$ is the common angle of the three σ bonds.[5]

2.2   Characteristics of the Fullerene Molecules

General Characteristics. The fullerene structure is unique in the sense that the molecules are edgeless, chargeless, and have no boundaries, no dangling bonds, and no unpaired electron. These characteristics set the fullerenes apart from other crystalline structures such as graphite or diamond which have edges with dangling bonds and electrical charges.[2]

Such features allow these molecules, and particularly the C_{60} which is the most symmetrical, to spin with essentially no restraint at a very high rate.[3]

The Fullerene Family. Many fullerene structures are theoretically possible, some with hundreds of atoms. Five have been unambiguously identified and structurally characterized so far: the C_{60}, C_{70}, C_{76}, C_{78}, and C_{84}.[7]

- At one end of the fullerene family is the C_{20}, the simplest structure, which is composed of twelve pentagons with no hexagon and with a pyramidization angle of 18°. All its atoms would have the sp^3 configuration but it is thermodynamically unstable.[8]

- The first stable fullerene and the first to be discovered is the C_{60}. It is the dominant molecule with sixty carbon atoms arranged so that they form twenty hexagons, in addition to the necessary twelve pentagons, giving it the appearance of a soccer ball (known as a football outside the U.S.A. and U.K.). Each hexagon is connected to alternating pentagon and hexagon, and each carbon atom is shared by one pentagon and two hexagons. Its hybridization is partially sp^3 and partially sp^2, with a pyramidization angle of 11.6°. It is shown schematically in Fig. 15.4. The C_{60} has a calculated diameter of 0.710 +/- 0.007 nm. It is intensely purple in apolar solvents.[7][8]

- The higher fullerenes are shown schematically in Fig. 15.5.[7] The C_{70} is the second most prominent fullerene. It has a rugby-ball appearance and is has a deep orange-red color in solution. The C_{76} is *chiral* with a structure consisting of a spiraling arrangement of pentagons and hexagons.[9] It has a bright yellow-green color in solution and in the crystal. The C_{78} is chestnut brown or golden-yellow and the C_{84} is olive-green. The electronic absorption spectra for these fullerenes is shown in Fig. 15.6.[8]

- As the number of carbon atoms increases, the s-character in the π-orbital becomes less and less pronounced to become zero in the case of the ideal graphite crystal, when the pyramidization angle is zero and the number of atoms is infinite.

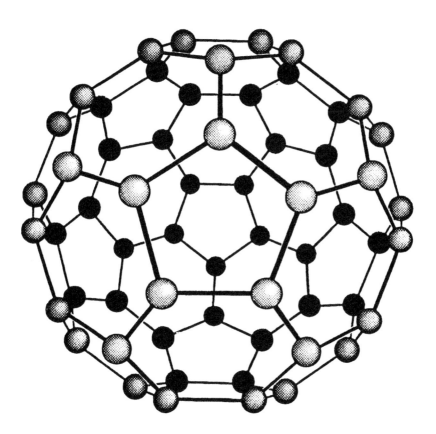

Figure 15.4. Schematic of a C_{60} fullerene molecule.

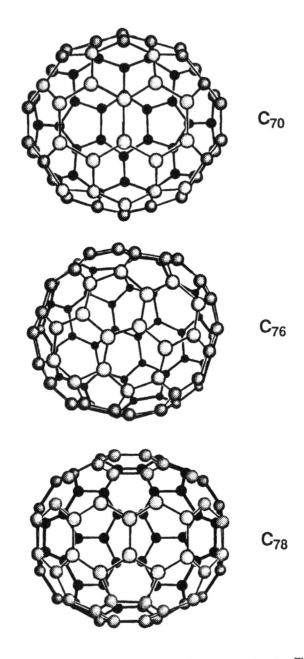

Figure 15.5. Schematic of C_{70}, C_{76}, and C_{78} fullerene molecules.[7]

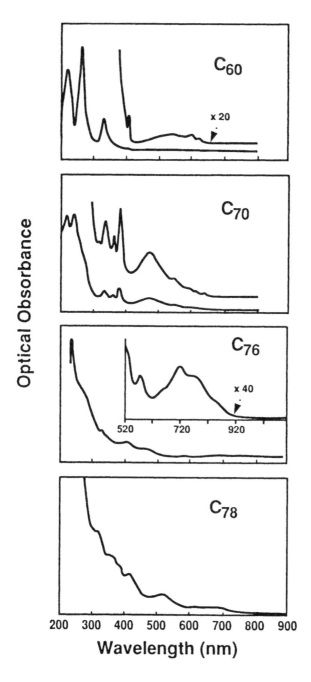

Figure 15.6. Electronic-absorption spectra of fullerenes in solution (C_{60} and C_{70} in hexane, C_{76} and C_{78} in dichloromethane).[7]

Related Structures. It appears that the spheroid fullerenes such as C_{60} are not the only type of large curved carbon molecules. Geometrically, many such structures are feasible.[10] Iijima and coworkers have actually detected carbon molecular structures by transmission electron microscopy that are in the form of cylindrical tubes, closed by polyhedral caps and known as "buckytubes". These structures have the hexagons arranged in a helix around the tube axis and some have a negative curvature.[11] The defect in the hexagonal network that triggers the formation of such structures is believed to be a single heptagonal ring. Some of these tubes are whisker-like and may reach micron size.

2.3   Mechanism of Formation

As described by Smalley, the formation of a fullerene follows an "efficient mechanism of self-assembly of an architecturally useful structure on a nanometer scale."[2]

The Isolated Pentagon Rule. A model of the growth mechanism of the fullerenes is based on the Isolated Pentagon Rule which states that such geometrical structure must meet the following three criteria: *(a)* it must be made only of hexagons and pentagons, *(b)* it must have twelve pentagons, and *(c)* it must not include adjacent pentagons.[2][7][12]

This last criterion is also required from the thermodynamics standpoint since the fusion of two pentagons is not favorable energetically due to increased ring strain, and carbon structures with adjacent pentagons are unstable.[7] The C_{60} is the first fullerene that avoids abutting pentagons. It is the most stable and symmetrical of all the stable fullerene molecules. Smaller geodesic structures, such as the C_{20} mentioned above, cannot form or be stable since they would incorporate abutting pentagons.

Formation. It had been originally suggested that the molecule begins as a graphitic network that adds carbon atoms as pentagons (in addition to the graphitic hexagons) causing the sheet to curl and eventually close to form a geodesic sphere. A more likely mechanism is formation by the aggregation of small carbon radicals.[2]   This occurs when graphite is vaporized at very high temperature. The condensing free-flowing graphitic sheets in the vapor have no atoms to form dangling bonds and the physical tendency to reach the lowest energy level induces them to curl and anneal into the fullerene structure.[3][13] A hypothetical growth sequence is shown in Fig. 15.7.[3]

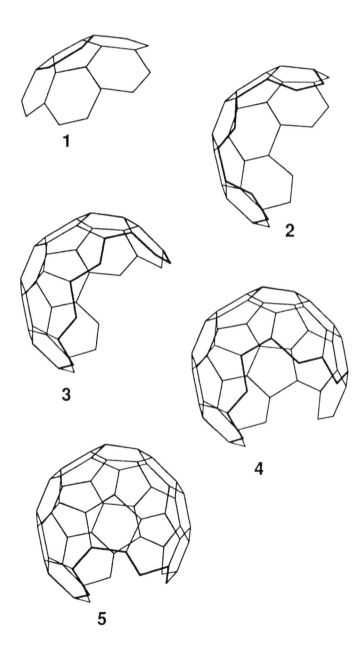

Figure 15.7. Hypothetical growth sequence of a C_{60} molecule.[3]

3.0   FULLERENES IN THE CONDENSED STATE

In the previous section, the formation and characteristics of single fullerene molecules were reviewed. In this and the following sections, the formation, characteristics, and properties of solid aggregates of fullerenes are examined. Fullerenes aggregates are a new discovery; their characterization and the determination of their properties is still at an early stage and much work remains to be done. Yet what has been accomplished so far shows the striking potential of these materials.

3.1   Crystal Structure of Fullerenes Aggregates

It was originally assumed that the solid aggregate of C_{60} fullerenes had a hexagonal close-packed structure. However, recent x-ray diffraction studies have shown unambiguously that it adopts the face-centered cubic (fcc) structure (providing that all solvent molecules are eliminated). Synchroton x-ray powder profile gives a lattice constant a = 1.417 nm. This value implies the close packing of pseudospheres having a diameter of 1.002 nm. This is consistent with the fitted radius of the C_{60} skeleton of 0.353 nm and a carbon van der Waals diameter of 0.294 nm, which is slightly smaller than that of graphite (0.335 nm). The intermolecular bonding is dominated by van der Waals forces, as confirmed by measurements of the isothermal compressibility.[14] The C_{60} aggregates, grown from solution, are shiny and black and reach 300 μm in size. They display a tenfold symmetry.[15]

The C_{70} aggregates are different as they have a hexagonal structure with lattice parameters a = 1.063 nm and c = 1.739 nm.[14]

3.2   Properties of Fullerenes Aggregates

Mechanical Properties. C_{60} aggregates are considered the softest of the solid phases of carbon.[14] However, calculations show that, under high pressure when compressed to less than 70% of its original volume, they could be harder than diamond.[3] They have high impact strength and resilience. They should also have high lubricity since the molecules are bonded by van der Waals forces in all planes which should allow the molecules to slip readily over each other in a manner similar to the *ab* planes of the graphite crystal.

The aggregates have a constricted micropore structure with a micropore width of 1.19 - 1.26 nm and a relatively high internal surface area (131.9 m^2/g).[16]

Semiconductor Properties. Calculations indicate that C_{60} could be a direct band-gap semiconductor similar to gallium arsenide. However, the semiconductor properties have yet to be determined.

Fullerene-Diamond Transformation. The rapid compression of C_{60} powder, to more than 150 atm in less than a second, caused a collapse of the fullerenes and the formation of a shining and transparent material which was identified as a polycrystalline diamond in an amorphous carbon matrix.[17] Thus the fullerenes are the first known phase of carbon that transforms into diamond at room temperature. Graphite also transforms into diamond but only at high temperatures and pressures (see Ch. 12, Sec. 3.0).

4.0 CHEMICAL REACTIVITY AND FULLERENE COMPOUNDS

4.1 Chemical Reactivity

Most of the studies of the chemical reactivity of the fullerenes have been done with C_{60} aggregates. Although the molecule is stable from the physical standpoint, it has a high electron affinity and is reactive chemically, especially with free radicals.[18]-[20]

Fullerenes are aromatic structures and dissolve readily in the archetypal aromatic compound, i.e., benzene and in other aromatic solvents. They oxidize slowly in a mixture of concentrated sulfuric and nitric acids at temperatures above 50°C. In pure oxygen, C_{60} begins to sublime at 350°C and ignites at 365°C; in air, it oxidizes rapidly to CO and CO_2 and is more reactive than carbon black or any other form of graphite.[16]

4.2 Fullerene Derivatives

A systematic approach to the chemistry of fullerene-organic compounds is beginning to emerge. Because of the unique character of the fullerenes, this chemistry is basically different from classical organic chemistry and these molecules may become the parents of an entirely new class of organic compounds.[21][22]

The C_{60} has been described as a "radical sponge" since it can be considered as having thirty carbon-carbon double bonds, where free radicals can attach themselves covalently on the outside of the carbon framework without destroying the kinetic and thermodynamic stability of the molecule. The stable compounds that are thus formed are called "exohedral", that is, formed outside the C_{60} shell.

Fullerene-Organic Compounds. Exohedral organic compounds investigated so far include the C_{60}-benzyl radicals ($C_6H_5CH_2$-), the C_{60}-allyl group (CH_2:$CHCH_2$-), and the C_{60}-methyl group (CH_3-).[19][20]

Fullerene-Organometallics. Also recently investigated are the organo-metallic exohedral complexes of osmium, ruthenium, and platinum which are readily attached to the external framework of the C_{60} molecule by solution chemistry.[21][23] An osmylated-C_{60} compound is shown in Fig. 15.8.[21]

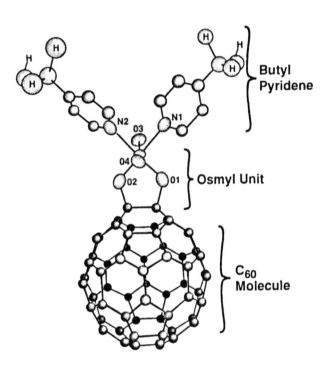

Figure 15.8. ORTEP drawing of the osmium tetraoxide adduct: $C_{60}(OsO_4)$(4-tert-butyl pyridine)$_2$.[21]

Fullerene-Fluorine Compounds. Fullerene-fluorine exohedral compounds are produced readily in the solid state when C_{60} aggregates are fluorinated at moderate temperatures (RT to $90^\circ C$ for several days). The material is covalently bonded with the composition $C_{60}F_{40}$ and a lattice parameter a = 1.705 - 1.754 nm.[24] Similar compounds are produced with the C_{70} molecule.[14]

4.3 Fullerene Intercalation Compounds

As mentioned above, fullerene aggregates are bonded by van der Waals forces so foreign elements are readily intercalated in the lattice in a manner similar to the intercalation of graphite reviewed in Ch. 10, Sec. 3.0. Intercalation elements investigated so far include the alkali ions (cesium, rubidium, potassium, sodium, and lithium) which, being smaller than the fullerene, fit into the lattice without disrupting the geodesic network and the contact between the molecules of the aggregate. A schematic of the structure of a M_6C_{60} compound is shown in Fig. 15.9.[14]

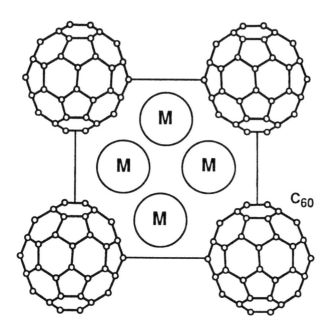

Figure 15.9. Schematic of M_6C_{60} compound. The metal can be K (lattice constant = 1.139 nm), Rb (1.154 nm), or Ca (1.179 nm). Small circles represents carbon atoms (not to scale).[14]

Superconductivity. Some intercalation compounds are superconductors, such as $C_{60}K_3$, with an onset of superconductivity at 17 K and zero
resistance at 5 K.[5][25]   Other compounds have even higher critical
temperatures, i.e., 30 K for rubidium-doped C_{60} and 43 K for rubidium-
thallium-doped C_{60}.[3]

4.4 Fullerene Endohedral Compounds

Different from the exohedral materials described in the preceding
section are the "endohedral" compounds, where the foreign atoms are
located within the fullerene cage instead of outside. These compounds are
designated $(M@C_n)$, the symbol "@" indicating that the M atom is located
within the C_n cage, "n" representing the number of carbon atoms of the
fullerene (60, 70, 76, etc.). Composition such as $(Y@C_{60})$ and $(Y_2@C_{82})$
(as identified by spectral analysis) are readily produced by laser vaporization of a graphite-yttria rod; other endohedral compounds with lanthanum
and uranium have also been synthesized.[2][3][18]

5.0   FULLERENES PROCESSING

The C_{60} and the higher fullerenes are produced from condensing
carbon vapors, providing that the condensation is sufficiently slow and the
temperature is sufficiently high ($>2000°C$).[2] These conditions exist in the
basic apparatus shown in Fig. 15.10.[26] An arc is generated between two
rod-electrodes made of high-purity graphite, in an atmosphere of pure
helium. The distance between the electrode tips is maintained constant.
The fullerenes are collected by solvent extraction of the resulting soot, with
a solvent such as N-methyl-2-pyrrolidinone. Yield of dissolvable fullerenes
as high as 94% are reported.[27]

In such a system, hydrogen, water vapor, and other contaminants
must be totally excluded since they would tend to form dangling bonds and
prevent closure of the fullerene molecule.[3] The helium atmosphere is
necessary because it slows migration of the carbon chains away from the
graphite rod and gives them sufficient time to form the initial cluster radicals
to rearrange into pentagons before the radicals start to polymerize.[13]

High fullerene yields are also reported in a benzene-oxygen flame with
optimum conditions of 9 kPa, a C/O ratio of 0.989 and a dilution of 25%
helium.[28]

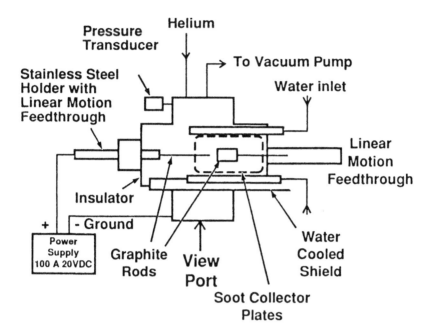

Figure 15.10. Schematic of apparatus for the production of fullerenes from graphite rods.[26]

6.0 POTENTIAL APPLICATIONS

It should be stressed that, at this stage, the applications of fullerenes are still on the drawing board and it will some time before they become a reality. The following potential applications have been suggested.[3]

- Direct band-gap semiconductor
- Superconductor (doped with potassium or rubidium)
- Non-metallic ferromagnetic material
- Gas storage and gas separation
- Purification of natural gas
- Fuel cells and hydrogen storage
- Storage for radioactive isotopes
- Lubricants

REFERENCES

1. Kroto, H. W., Heath, J. R., O'Brien, S. C., Curl, R. F., and Smalley, R. E., *Nature*, 318:162-163 (1985)

2. Smalley, R. E., *Acc. Chem. Res.*, 25:98-105 (1992)

3. Curl, R. F. and Smalley, R. E., *Scientific American*, 54-63 (Oct. 1991)

4. Krätschmer, W., Lamb, L. D., Fostiropolous, K., and Huffman, D. R., *Nature*, 347:354-358 (1990)

5. Haddon, R. C., *Chem. Res.*, 25:127-133 (1992)

6. Johnson, R. D., Bethune, D. S., and Yannoni, C. S., *Acc. Chem. Res.*, 25:169-175 (1992)

7. Diederich, F. and Whetten, R. L., *Acc. Chem. Res.*, 25:119-126 (1992)

8. Nuñez-Regueiro, M., *La Recherche*, 23:762-764 (June 1992)

9. Etti, R., Diederich, F., and Whetten, R. L., *Nature*, 353:149-153 (1991)

10. Terrones, H. and Mackay, A. L., *Carbon*, 30(8):12251-1260 (1992)

11. Iijima, S., Ichihashi, T., and Ando, Y., *Nature*, 356:776-778 (30 April 1992)

12. Kroto, H. W. and McKay, K., *Nature*, 331:328-331 (28 Jan. 1988)

13. Pang, L. S. K., et al., *Carbon*, 30(7) (1992)

14. Fischer, J. E., Heiny, P. A., and Smith A. B., *Acc. Chem. Res.*, 25:112-118 (1992)

15. Ceolin, R., et al., *Carbon*, 30(7):1121-1122 (1992)

16. Ismail, I. M. K., *Carbon*, 30(2):229-239 (1992)

17. Nuñez-Regueiro, M., et al., *Nature*, 355:237 (1992)

18. McElvany, S. W., Ross, M. M., and Callahan, J. H., *Acc. Chem. Res.*, 25:162-168 (1992)

19. Wudl, F., *Acc. Chem. Res.*, 25:157-161 (1992)

20. Baum, R. M., *C&EN*, 17-20 (Dec. 16 1991)

21. Hawkins, J. M., *Acc. Chem. Res.*, 25:150-156 (1992)

22. Olah, G. A., et al., *Carbon*, 30(8):1203-1211 (1992)

23. Fagan, P. J., Calabrese, J. C., and Malone, B., *Acc. Chem. Res.*, 25:134-142 (1992)

24. Nakajima, T. and Matsuo, Y., *Carbon*, 30(7):1119-1120 (1992)

25. Hebard, A. F., et al., *Nature*, 350:600-601 (April 18, 1991)

26 Parker, D. H., et al., *J. Am. Chem. Soc.*, 113:7499-7503 (1991)

27. Parker, D. H., et al., *Carbon*, 30(3):1167-1182 (1992)

28. Howard, J. B., et al., *Carbon*, 30(8):1183-1201 (1992)

Glossary

Ablation: The removal of material from the surface of a body exposed to a high-velocity gas such as a reentry nose cone or a rocket motor. Ablation occurs mainly by decomposition or vaporization resulting from the friction with the gas and the resulting high temperature.

Activation: A process that increases the surface area of a material such as charcoal or alumina. In the case of charcoal, the surface of the material is oxidized and minute cavities are created which are capable of *adsorbing* gas atoms or molecules.

Adsorption: The formation of a layer of gas on the surface of a solid (or occasionally a liquid). The two types of adsorption are *(a)* chemisorption where the bond between the surface and the attached atoms, ions, or molecules is chemical, and *(b)* physisorption where the bond is due to van der Waals' forces.

Aliphatic Hydrocarbons: A group of organic compounds having an open-chain structure such as parafin, olefin, acetylene, and their derivatives.

Angular Momentum: The product of moment of inertia and angular velocity.

Aqua Regia: A mixture of concentrated nitric acid and concentrated hydrochloric acid in the ratio 1:3 respectively.

Aromatic Hydrocarbons: A group of organic compounds that contain a benzene ring in their molecules or that have chemical properties similar to benzene.

Atomic Mass Unit (amu): The atomic mass unit is defined as 1/12th of the atomic mass of carbon-12, the most common isotope of carbon. There are 1.660 33 x 10^{24} amu per g (see **Avogadro's Constant**).

Atomic Number: The number of protons in the nucleus of an atom which is equal to the number of electrons revolving around the nucleus. The atomic number determines the location of the element in the Periodic Table.

Avogadro Constant (N_A): (Formally Avogadro's number) The number of atoms or molecules contained in one mole of any substance (6.0221367 x 10^{23}). Avogadro proposed that each distinct substance in the gaseous state consists of characteristic discrete particles called molecules: a specified volume of any gas measured at a uniform pressure and temperature contains the same number of molecules.

Bandgap: The energy-distance electrons have to move to go from the valence band to the conductor band.

Benzene Ring: The structure of the hydrocarbon compound benzene, C_6H_6. It is a six-carbon ring where all C-C bonds are equivalent and intermediate in length between single and double bonds. One electron per carbon atom (for a total of six) is delocalized. These six electrons have equal probability of being found anywhere around the ring. They reside in pi bonds which are perpendicular to the plane of the molecule. Benzene is the archetypal aromatic compound.

Bort: An imperfectly crystallized form of natural diamond or diamond fragment, used mostly as an abrasive.

Breccia: A sedimentary rock composed mainly of large angular mineral fragments embedded in a fine-grained matrix. The particles are usually derived from the same parent formation.

Carbene: A species of the type R_2C: in which two electrons of the carbon atom do not form bonds. Carbenes are very reactive and exist only as transient intermediate compounds. A typical example is methylene, $:CH_2$.

Carbonado: An impure, dark-colored, non-transparent aggregate of polycrystalline diamond characterized by its great toughness.

Cathodoluminescence: The property of emitting light when a material is bombarded by electrons.

Chirac Structure: A structure that exists in left-and right-handed forms.

Coefficient of Friction: The ratio of the force required to move one surface over another to the total force pressing the two surfaces together.

Coefficient of Thermal Expansion (CTE): The fractional rate of increase in linear dimension as a function of temperature at a constant pressure.

Condensation Reaction: A chemical reaction in which two molecules combine to form a larger molecule. Condensation is accompanied by the elimination of a small molecule, i.e., water.

Covalent Bond: A primary force of strong attraction holding atoms together in a molecule or crystal in which pairs of valence electrons are shared. The covalent bonds in which one pair of electrons is shared are known as single bonds. Two pairs of shared electrons form a double bond and three pairs a triple bond.

Dielectric Constant: The ratio of electric displacement in a dielectric medium to the applied electric-field strength, defined by "e" in the following equation:

$$F = QQ'/er^2$$

where F is the force of attraction between two charges Q and Q´ separated by a distance r in a uniform medium.

Direct-Bandgap Semiconductor: see Indirect-Bandgap Semiconductor.

Dubinin Equation: The expression of Dubinin's theory for the filling of micropores (see Ch. 10, Ref. 16).

Electromagnetic Interference (EMI): A generally undesirable "noise" in the form of magnetic or electrical energy, mostly occurring the radio-frequency portion of the electromagnetic spectrum (1 kHz to 10 GHz). Sources of EMI are lightning, computers, communication systems, and many others.

Electron Volt (eV): The unit of energy accumulated by a particle with one unit of electrical charge while passing through a potential difference of one volt. It is used to measure the energy of particles but is not a SI unit. 1eV = 1.602×10^{-19} Joule.

Epitaxy: The growth on a crystalline substrate of a crystalline substance that has essentially the same crystal parameters as the substrate. The substance can be grown from a gas (gas-phase epitaxy) or from a liquid (liquid-phase epitaxy).

Galvanic Corrosion: Accelerated corrosion of a metal due to an electrical contact with a more noble metal or non-metallic conductor.

Galvanic Couple: A pair of dissimilar conductors, commonly metal, in electrical contact.

Gas-Phase Epitaxy: See **Epitaxy.**

Ground State: The state of an atom at its minimum energy level. The electrons are as close to the nucleus as they can be. In the ground state, the atom cannot radiate energy. A number of possible orbits extend outward from the nucleus where electrons can be lifted to form the excited states. Any electron can be lifted from the ground state to the excited state by absorbing a definite amount of energy.

Igneous Rock: A rock formed by solidification from the molten state (molten magma).

Index of Refraction: The ratio of the velocity of light in air to the velocity of light in a substance at a given wavelength.

Indirect-Bandgap Semiconductor: A semiconductor material such as silicon in which a valence electron with energy E_g cannot be excited directly across the forbidden band between the valence and conduction bands (the energy gap being E_g), but requires a change in momentum. Conversely, in a direct-bandgap semiconductor such as gallium arsenide, the electron can make the transition directly. It does so by absorbing or emitting a photon of energy E_g.

Ionic Bond: Atomic bonding due to electrostatic attraction between charged particles (ions) resulting from the transfer of electrons.

Ionization Potential: The minimum energy necessary to remove an electron from an atom or molecule to such a distance that no electrostatic interaction remains between the electron and the ion. Synonym: ionization energy. Units: electron volt or Joule per mole.

Isomers: Molecules with the same number of atoms of the same elements but with different structural arrangements and properties. *Isomerization* is the process that causes a substance to be changed into an isomer.

Isotope: One of two or more atoms of the same element with the same number of protons in the nucleus but different number of neutrons.

Kimberlite: An igneous rock *(peridotite)* containing garnet and olivine which probably originated in the upper mantle and was forced upwards in volcanic pipes.

Lonsdaleite: A naturally occurring hexagonal form of diamond sometimes found in meteorites.

Mafic Mineral: Dark-colored mineral rich in iron and magnesium, such as pyroxene, amphibole, or olivine.

Metallic Bond: The interatomic chemical bond in metals characterized by delocalized electrons in the energy bands. The atoms are considered to be ionized with the positive ions occupying the lattice positions. The valence electrons are free to move. The bonding force is the electrostatic attraction between ions and electrons.

Metamorphic Rock: A rock that has been submitted to changes in mineralogy and texture by pressure and temperature in the earth's interior.

Molar Density: Total number of atoms per cm^3 divided by Avogadro's number.

Molecular-Beam Epitaxy: A deposition process based on evaporation capable of producing high-purity thin films with abrupt composition changes.

Molecule: A group of atoms held together by covalent or coordinate bonds (found in most covalent compounds). Ionic compounds do not have single molecule since they are a collection of ions with opposite charges. A diatomic molecule is formed by two atoms (e.g., H_2, HCl) and a polyatomic molecule by several atoms (e.g., carbon C_3, C_4, etc.)

Node: A point or region in a standing wave in which a given characteristic of the wave motion, such as particle velocity or displacement, or pressure amplitude, has a minimum or zero value.

Orbital: A region in which an electron may be found in an atom or molecule (see **Standing Wave**).

Pack Cementation: A coating process relying on chemical transport in a closed system.

Pauli's Exclusion Principle: The principle that no two identical elementary particles having half-integer spin (fermion) in any system can be in the same quantum state (i.e., have the same set of quantum numbers). In order to account for the various spectral characteristics of the different elements, one must assume that no two electrons in a given atom can have all four quantum numbers identical. This means that, in any orbit (circular, elliptical, or tilted), two electrons at most may be present; and of these two, one must spin clockwise and the other must spin counterclockwise. Thus, the presence of two electrons of opposite spin in a given orbit excludes other electrons.

Peridotite: A coarse-grained mafic igneous rock composed of olivine with accessory amounts of pyroxene and amphibole but little or no feldspar.

Pi (π) Bond: The bond resulting from the pi orbital, this orbital being a molecular orbital produced by sideways overlap of the p-orbital, such as in a carbon molecule.

Plumbago: An old term for graphite.

Polymer: A substance having large molecules consisting of repeating units (the monomers).

Polymorph: One of several structures of a chemical substance. Graphite and diamond are polymorphs of carbon.

Positron: An elementary particle with the mass of the electron and a positive charge equal to the negative charge of the electron. It is the antiparticle of the electron.

Pyrolysis: Chemical decomposition occurring as a result of high temperature.

Quantum Numbers: The numbers characterizing a region in which an electron may move. These numbers are "n", the principal quantum number (ground state is n = 1, exited states have n = 2, 3 or more), "l", the orbital quantum number, "m", the magnetic quantum number, and "s", the spin quantum number (1/2 and -1/2) (see **Pauli's Exclusion Principle**).

Radiation: Energy travelling in the form of electromagnetic waves or photons, or a stream of particles such as alpha- and beta-particles from a radioactive source or neutron from a nuclear reactor.

Radiation Energy: The energy of the quantum of radiation hv, being equal to the energy difference of the states. No electron radiates energy as long as it remains in one of the orbital energy states; energy occurs only when an electron goes form a higher energy state to a lower one.

Resilience: The amount of potential energy stored in an elastic substance by means of elastic deformation. It is usually defined as the work required to deform an elastic body to the elastic limit divided by the volume of the body.

Sigma (σ) Bond: The bond resulting from the sigma orbital, this orbital being a molecular orbital produced by overlap along the line of axes, such as in a carbon molecule.

Spin: The intrisic angular momentum of an elementary particle or group of particles. Spin is characterized by the *quantum number* "s". Because of their spin, particles also have their own intrinsic magnetic moments.

Standing Wave (or Stationary Wave): A wave incident on the boundary of the transmitting medium, reflected either wholly or partially according to the boundary conditions. The reflected wave is superimposed on the incident wave resulting in an interference pattern of nodes and antinodes. The motion of the electron can be considered as a standing wave and can be expressed by a differential equation (for the hydrogen atom only). The solutions (wave functions) determine the coordinates of the electron, the nucleus being the origin of the coordinate system. The wave functions are known as orbitals.

Stefan-Boltzmann Law: The law relating the radiant flux per unit area emitted by a black body to the temperature, with the formula $M_c = \sigma T^4$, where M_c is the radiant flux leaving a surface per unit area and σ is the Stefan-Boltzmann constant.

Stereospecific: Describing chemical reactions that give products with a particular arrangement of atoms in space.

Sub-Shell: A division of an electron shell. The electron shell of principal quantum number n can be divided into n sub-shells. The first has two electrons and subsequent ones can contain four more than the preceding one (6, 10, 14, 18, etc.). The sub-shells are symbolized as s, p, d, f, g, h and i (these letters being derived from atomic spectroscopy nomenclature). The first electron shell has only the 1s sub-shell, the second contains a 2s subshell and a 2p subshell, etc.

Temperature Coefficient of Resistance: The change (usually small) in the resistance of a material as a function of the changes of its thermodynamic temperature.

III - V, II - VI Semiconductors: Compounds with semiconducting properties consisting of elements of Groups III and V such as gallium arsenide or
elements of Groups II and VI such as zinc selenide.

Tow: A loose, essentially untwisted strand of synthetic fibers.

Ultramafic Rock: An igneous rock consisting dominantly of mafic minerals, containing less than 10% feldspar and including dunite, peridorite,
amphibolite, and pyroxenite.

Valence Electron: An electron located in the outer shell of an atom which
is able to participate in the formation of chemical bonds.

Van der Waals' Forces: Interatomic and intermolecular forces of electrostatic origin. These forces arise due to the small instantaneous dipole
moments of the atoms. They are much weaker than valence-bond forces
and inversely proportional to the seventh power of the distance between the
particles (atoms or molecules).

Vapor: A gas at a temperature below the critical temperature so that it can
be liquefied by compression without lowering the temperature.

Wave Function: A mathematical expression representing the coordinates
of a particle (i.e., an electron) in space. It appears in the Schroedinger
equation in wave mechanics.

Young's Modulus: The ratio of the applied load per unit area of cross
section to the increase in length per unit length of a body obeying Hooke's
law.

REFERENCES

Dictionary of Chemistry, (J. Daintith, ed.), Oxford University Press, Oxford, UK (1992)

Dictionary of Physics, (V. Illingworth, ed.), Penguin Books, London, UK (1991)

Cram, D. J. and Hammond, G. S., *Organic Chemistry*, McGraw-Hill, New York (1964)

Eggers, D. F., Jr., et al., *Physical Chemistry*, John Wiley & Sons, New York (1964)

Press, F. and Siever, R., *The Earth*, W. H. Freeman & Co., San Francisco, CA (1974)

Index

Printed and bound by CPI Group (UK) Ltd, Croydon, CR0 4YY

03/10/2024

01040434-0009